Captivity Literature and the Environment

In his study of captivity narratives, Kyhl Lyndgaard argues that these accounts have influenced land-use policy and environmental attitudes at the same time that they reveal the complex relationship between ethnicity, landscape, and authorship. In connecting these themes, Lyndgaard offers readers an alternative environmental literature, one that is dependent on an understanding of nature as home rather than as a place of temporary retreat. He examines three captivity narratives written in the 1820s and 1830s—A Narrative of the Life of Mrs. Mary Jemison, The Captivity and Adventures of John Tanner, and Life of Black Hawk—all of which engage with the Jacksonian policy of Indian removal and resist tropes of the so-called Vanishing Indian. As Lyndgaard shows, the authors and the editors with whom they collaborated often saw their stories as a plea for environmental and social justice. At the same time, audiences have embraced them for their vision of a more inclusive and less exploitative American society than was proffered by the rhetoric of Manifest Destiny. Their legacy is that while environmental and social justice has been slow in fulfilment, their continued popularity testifies to the fact that the struggle for justice has never been ceded.

Kyhl D. Lyndgaard is Director of First-Year Seminar and the Writing Center at the College of Saint Benedict and Saint John's University, Collegeville, MN, USA.

North American Literature and the Environment, 1600–1900

Series editor: Matthew Wynn Sivils, Iowa State University, USA

Captivity Literature and the Environment
Nineteenth-century American cross-cultural collaborations
Kyhl D. Lyndgaard

Captivity Literature and the Environment

Nineteenth-century American cross-cultural collaborations

Kyhl D. Lyndgaard

Routledge
Taylor & Francis Group
LONDON AND NEW YORK

First published 2017
by Routledge

2 Park Square, Milton Park, Abingdon, Oxfordshire OX14 4RN
52 Vanderbilt Avenue, New York, NY 10017

Routledge is an imprint of the Taylor & Francis Group, an informa business

First issued in paperback 2018

British Library Cataloguing in Publication Data
A catalogue record for this book is available from the British Library

Library of Congress Cataloging-in-Publication Data

Names: Lyndgaard, Kyhl, author.
Title: Captivity literature and the environment: nineteenth-century American cross-cultural collaborations / by Kyhl D. Lyndgaard.
Description: New York: Routledge, 2017. | Series: North American literature and the environment, 1600–1900 | Includes bibliographical references and index.
Identifiers: LCCN 2016015661
Subjects: LCSH: Captivity narratives—United States—History and criticism. | Captivity in literature. | American literature—19th century—History and criticism. | Environmental literature—United States—History and criticism.
Classification: LCC PS169.C36 L96 2017 | DDC 810.9/35873—dc23
LC record available at https://lccn.loc.gov/2016015661

ISBN: 978-1-4724-8519-9 (hbk)
ISBN: 978-0-367-14043-4 (pbk)

Typeset in Sabon
by codeMantra

Contents

List of figures and table

Figures

Table

List of appendices

Publication histories of three captivity narratives

Acknowledgments

The origins of this book can be traced back to a seminar offered as part of the Graduate Program in Literature and Environment at the University of Nevada, Reno. The seminar was taught by Michael P. Branch, whose own work tracing environmental traditions in literature offered an invaluable model. Mike gave me unstintingly generous encouragement and help throughout the process of writing and revising the entire manuscript over a period of many years.

At Nevada, Cheryll Glotfelty also played a key role in convincing me that this project was a book, while Bill Rowley, Marybeth Eleanor Nevins, and Kathleen Boardman thoroughly commented on the work. Others in the Literature and Environment program, particularly Jim Bishop and Dave Johnson, offered much support and conversation, both on campus and while hiking and mountain biking in the Great Basin and Sierras.

I thank the library staff at UNR, as well as at Marlboro College and at the College of Saint Benedict/Saint John's University, for all their help in fulfilling dozens of interlibrary loan requests. My colleagues at Marlboro were supportive of my research in many ways, and I want to specifically thank John Sheehy, Gloria Biamonte, and Richard Glejzer. My work on this book has been informed and shaped by conversations and correspondence with members of the Association for the Study of Literature and Environment.

At Ashgate and Routledge, I am indebted to Matthew Wynn Sivils and Ann Donahue for their initial work on the project, and to Michael Bourne and Elizabeth Levine in the final stretch; I thank the entire publishing staff for a smooth process.

An earlier and somewhat more brief version of Chapter 3, "Scientific and Sympathetic Collaboration: Edwin James and John Tanner," appeared previously. "Landscapes of Removal and Resistance: Edwin James's Nineteenth-Century Cross-Cultural Collaboration," by Kyhl Lyndgaard is reproduced from *Great Plains Quarterly* with permission from the University of Nebraska Press. Copyright 2010.

For their interest and enthusiasm for my work, I thank my parents, Dave and Barb Lyndgaard. And for the loving ability to show excitement and patience in tandem, I thank my wife, Marian. To her, and to our captivating children Lars and Peregrine, I dedicate this book.

Acknowledgements

Prologue

Taking off the moccasin flower and putting on the lady's slipper: Indian removal and the natural environment in the nineteenth century

One year before the passage of the Indian Removal Act in 1830, Almira Lincoln Phelps published the first edition of her *Familiar Lectures of Botany*. Discussing native orchids of America, including what was commonly known as the moccasin flower, she explained that these flowers, "particularly the Orchis tribe," which are "opposing all attempts at civilization, are to be found only in the depths of the forest [...]; we may, in this respect, compare them to the aboriginal inhabitants of America, who seem to prefer their own native wilds to the refinements and luxuries of civilized life" (175–76). Throughout the nineteenth century, the moccasin flower served as a synecdoche for Native Americans in intense national debates over federal policies of Indian removal. These distinctive orchids, with pouch-like flowers, are members of the genus *Cypripedium* and occur naturally throughout the Northern Hemisphere. In North America, many different species were commonly called moccasin flowers at the beginning of the nineteenth century. Yet by 1900, the most common term used was the same as that used in England: the lady's slipper. This shift in terminology ominously mirrors changes in white American perceptions of the seeming necessity of Indian removal. The ways white Americans wrote about moccasin flowers—and the ways the earlier name itself slowly fell from common usage—illustrate the problematic ways that Indian removal was rationalized alongside anthropogenic changes in the landscape. A brief exploration of the changing discourse of moccasin flowers provides a helpful backdrop for the captivity narratives I analyze in this book.

Phelps links the moccasin flower to Native Americans, calling it a member of a "tribe," and then makes the connection even more explicit. She suggests that the plant cannot be civilized, leaving unclear the fate of the plant in the face of the onrush of white settlers. The story of the moccasin flower suggests how widely present the debate over removal was, having reached even botanical texts. And *Familiar Lectures in Botany* is an excellent example of the pervasiveness of such lessons, as it was widely read, remained in print for decades, and sold "more than 275,000 copies by 1872" (Gianquitto 22). The necessity for such a culturally embedded argument may be seen as evidence that Indian removal was not an inevitable conclusion, but rather a choice. In other words, allusions to Native

Americans in so unlikely a place as a botany textbook suggests that their removal westward was not a foregone conclusion, but rather something that needed to be taught to students. Phelps's stark language suggests that students left to think about Indian removal on their own might reasonably arrive at a different conclusion. Her textbook "offers students a model of right action for their behavior, particularly in the home" (Gianquitto 26). By learning what Phelps calls the "symbolical language of flowers," students also learned the moral orientation expected of them by nineteenth-century white American society, which included a belief in the trope of the "Vanishing Indian."

Lee Clark Mitchell, in *Witnesses to a Vanishing America: The Nineteenth-Century Response*, argues that the connections between environmental and social justice for Native Americans cannot be teased apart: "At every point—wilderness conservation on the one hand and tribal preservation on the other, appreciation for the land and respect for indigenous cultures, destruction of the environment and extinction of Indian tribes—the issues seem to fall into logical, even obvious associations" (263). I do not suggest that the destruction of moccasin flowers is equivalent to the destruction of human lives and cultures. However, on a symbolic level, the moccasin flower—in name, habitat, and cultural connotation—functions as a useful cultural shorthand for the complicated and intertwined history of Indian removal and changes in the North American landscape during the nineteenth century.

While the texts I later focus on provide examples of resistance to the Indian Removal Act, many captivity narratives—both nonfictional and fictional—suggest viewpoints similar to Phelps's. For example, James Fenimore Cooper's novel *The Last of the Mohicans*, which features various captivities, suggests a doomed and tragic trajectory for the Mohican tribe, however "noble" Chingachgook might be. Fenimore Cooper also uses flowers as stand-ins for women, specifically domestic flowers for white women. Hawk-eye says, "These Mohicans and I, will do what man's thoughts can invent, to keep such flowers, which, though so sweet, were never made for the wilderness, from harm" (46). Other texts were outright propaganda in favor of Indian removal, such as Archibald Loudon's *A Selection of Some of the Most Interesting Narratives of Outrage Committed by the Indians* (1808–11). Loudon's preface directly engages debates over Native Americans, noting that while the "philosopher [...] sees green fields and meadows in the customs and virtues of the savages. ... an uncivilized Indian is but a little way removed from a beast who, when incensed, can only tear and devour" (v). Clearly, debates over Indian removal were motivations for writing and publishing captivity narratives.

Phelps's world view generally aligned with federal policies of Indian removal. In her book's first and fifth editions (1829, 1836) she claimed that "the discoveries and observations [of the medicinal qualities of plants] of the Indians have perished with themselves" (first edition 13; fifth edition 14).

All references to moccasin flowers are purged from the 1836 edition. When discussing native orchids, however, Phelps adds this startlingly harsh line:

> It would seem too that the freaks of these vegetable beings are not designed for our observation, for they are as peculiar in their choice of habitations as in their external forms, preferring wildness, barrenness, and desolation to the fostering care of man, or the most luxuriant soil. (188)

The lady's slipper, according to her "Symbolical Language of Flowers," stands for "capricious beauty"—a type of beauty that Phelps would have found to be in conflict with the period's gender expectations (172). In these few years between editions, Phelps has already "taken off" the moccasin flower.

Phelps's use of the moccasin flower as a stand-in for Indians was hardly unique. Consider William Cullen Bryant, who used his editorship at the *New-York Evening Post* as a platform to editorialize in support of parks and conservation but also wrote in support of President Andrew Jackson's Indian Removal Act (Muller, 87–90).[1] Bryant, like fellow New Yorker Fenimore Cooper, saw Native Americans as belonging to a vanishing and singular culture. His poetry often drew upon flowers, including the moccasin flower, to symbolize not only death in general, but also the supposedly inevitable extinction of Native American culture in poems such as "The Prairies," "An Indian at the Burial-Place of his Fathers," "The Death of the Flowers," or "The Maiden's Sorrow." In the latter poem, published in 1842, Bryant writes,

> Far on the prairies of the west,
> None who loved thee beheld thee die;
> They who heaped the earth on thy breast
> Turned from the spot without a sigh.
> There, I think, on that lonely grave,
> Violets spring in the soft May shower;
> There, in the summer breezes wave
> Crimson phlox and moccasin flower. (64)

Bryant here suggests that Native Americans, already relegated in his mind to the "prairies of the west," are already sent to exile, unloved, and are thus doomed to vanish. Bryant's writing is full of contemporary references that his audience readily understood, as was Phelps's, and both were meant to instruct their readers in a worldview that was highly prescriptive of social roles in regard to gender and ethnicity. Moccasin flowers would not sprout from disturbed ground on the sunny prairie, as they require woodland and bog habitats. Many Native American communities, like many of the moccasin flower colonies, were severely disrupted by the changes in the land that white pioneers instituted as they pushed the frontier westward.

Echoing Bryant's imagery, but in strong opposition to his conclusions, botanist, explorer, and writer Edwin James lamented in an 1859 letter that

> [t]he locomotive engine and the railroad car scour the plain in place of the wolf and curlew. Mayweed and dog fennel, stink weed and mullein have taken the place of 'purple flox and the mocassin flower,' the Celt, the Dane, the Swede and the Dutchman are instead of Black Hawk and Wabashaw, Wabouse, Manny-Ozit and their bands" (qtd. in Pammel 177).

James, one of the subjects of the second chapter of this book, argued in the 1820s and 1830s for the preservation of both the natural environment and the Native American peoples of the Great Lakes and the Great Plains regions. That James, an accomplished botanist, renders the names of phlox and the moccasin flower within quotation marks—and with irregular spelling—suggests that he is referencing a common phrase of the day and perhaps is even citing Bryant. His writing clearly connects industrial development and westward expansion with the twin dislocation of native plants and Native Americans. The plants that he observes from his home in Iowa as growing from the disturbed prairie are invasive, exotic plants that were (and are) considered weeds.

A prolonged and valuable comparison between weeds and native flowers was made by Cooper's daughter, Susan Fenimore Cooper, in her 1850 book *Rural Hours*:

> The wild natives of the woods grow there willingly, while many strangers, brought originally from over the Ocean, steal gradually onward from the tilled fields and gardens, until at last they stand side by side upon the same bank, the European weed and the wild native flower.
>
> These foreign intruders are a bold and hardy race, driving away the prettier natives. It is frequently remarked by elderly persons familiar with the country, that our own wild flowers are very much less common than they were forty years since. … The strange pitcher-plant is said to have been much more common, and the moccasin-flower abounded formerly even within the limits of the village. Both are now rare, and it is considered a piece of good luck to find them. (49–50)

Cooper opens this passage by linking disturbed ground to the ability of weeds to take hold where native plants formerly made up the entire population. She specifies that the weeds are European, and constitute a "bold and hardy race" that continues to displace the native population. The moccasin flower is used as one of the primary examples of the decimated native flora. Despite the elegiac tone, Cooper claims that this is a "mingled society" in which both examples of both European and North

American origins can "stand side by side" (49). Despite the losses of the past forty years, she believes that both populations may yet live together. As Rochelle Johnson argues, Cooper was "troubled by how [the American picturesque] ... affected people's experience of the history of American landscapes" (71). The then-emergent idea of Manifest Destiny "silenced the natural world ... Cooper gave voice to those overlooked aspects of nature's story, thereby challenging her contemporaries' vision of nature as a metaphor for progress" (Johnson 71).

Another poet who used flowers as representative of whites and Native Americans in the 1800s was Lucy Larcom.[2] Best known for describing her experience of having to go to work in a textile mill at age eleven in her book *A New England Girlhood*, Larcom "respects the child's experience ... especially outdoor play and unmediated access to the natural world" (—Kilcup 180). "Manitou's Garden" is an 1854 poem written for children that describes how a "Chippewa boy" refuses an offer of flowers from "flaxen-haired Fred," saying, "Nor yet will [I] go / Where lilies and roses / Like pale captives grow" (80). The Chippewa boy suggests that the white boy come play with him instead, and notes that "The moccasin-flower / Springs up where [Manitou] stopped" (80). The poem ends in stalemate, as neither boy is willing to venture onto the other's playground. Importantly for my project, the poem connects domestic flowers with captivity, although in the poem this looming danger is avoided through the timely intervention of Fred's mother. Larcom also challenges the industrialization and urbanization of America, including a critique of both slavery and Indian removal. Despite Larcom's best efforts in her creative work, the cult of domesticity and true white womanhood offered no solution. Nonetheless, the suggestion that physical and cultural borders can be crossed by an individual demonstrates that the dichotomies of wilderness/barbaric and settled/civilized do not provide an accurate ordering of the world, regardless of the supposed inevitability of the displacement of Native American cultures. In poems such as "Flowers of the Fallow," Larcom further works to complicate the boundaries of wild/civilized and fallow/cultivated through specific species of flower such as mullein and yarrow.

Yet some white Americans did see alternatives to Indian removal. Turning again to the moccasin flower, some fascinating corollaries were advanced as the orchids became increasingly scarce as the nineteenth century progressed. Several articles promoted an assimilatory argument to ensure that the moccasin flower did not vanish. In 1863, a brief piece called "Domestication of Wild Flowers" in the *Maine Farmer* notes that the moccasin flower

> is very fine and desirable, and abounds along the margins of some of our swamps. [...] We have succeeded in removing successfully nearly all the wild plants in our vicinity that are desirable [...]. Far too little attention is paid to the wild flora of our country, and the beauties of

> our gardens would be much enhanced if more of them were removed to the borders" (Perdue 1).

A source of nationalist pride, moccasin flowers are "removed" from swamps and placed into civilized garden "borders." This pattern of removal is repeated throughout the middle and later nineteenth century. Connected with the near contemporary opening of some of the first Indian boarding schools (Carlisle Indian Industrial School, to name perhaps the most prominent, was founded in 1879), many white Americans believed Native Americans needed to be assimilated into white culture, forcibly if need be. Indeed, many boarding school experiences and the historical record show that captivity was suffered by Native Americans as well as whites.[3]

Even more striking passages are found throughout an 1891 article called "Cultivation of Native Orchids." Author Laura Sanford writes that

> [t]he time has come when, if we are to possess the representatives of this fair and fascinating race, which belong to us as the native offspring of American soil, we must guard them from destruction in their favorite haunts [...]. And we must not only endeavor so to guard them in their old-time nooks, but we must draw them within the shelter of cultivation (30).

Words such as "possess" obviously support captivity and domestication as a perceived necessity. The connection between race and flowers is again made, and a romantic pride in America's native populations is expressed using rhetoric not unlike Bryant's. The sentences quoted here, which are conspicuously devoid of words identifying the subject as a plant, could be plucked from their source and mistaken for the words of progressive Indian reformers of the latter half of the nineteenth century. Sanford ultimately asserts that the moccasin flower dies from "homesickness" after a year or two. As seen by death rates upon removal to reservations or boarding schools, Native Americans also suffered terribly upon the loss of their homes.

Elaine Goodale Eastman's fifty-two line long poem "Moccasin Flower," which Sanford knew, includes lines pointing to a more nuanced relationship among whites, Native Americans, and native orchids:

> Yet shy and proud among the forest flowers,
> In maiden solitude,
> Is one whose charm is never wholly ours,
> Nor yielded to our mood:
> One true-born blossom, native to our skies,
> We dare not claim as kin,
> Nor frankly seek, for all that in it lies,
> The Indian's moccasin. (46)

Eastman thus describes in code the tension between the various North American inhabitants. Published as part of her book *In Berkshire with the Wildflowers* (1879) when she was only sixteen, and written in collaboration with her younger sister, the poem clearly articulates the resistance of the "native," rather than relying upon the trope of vanishing. At this point in the poem Eastman pivots slightly, spending the next ten lines on the difficulties of finding the moccasin flower. She ends the poem on a strange note:

> With careless joy we thread the woodland ways
>   And reach her broad domain.
> Thro' sense of strength and beauty, free as air,
>   We feel our savage kin,—
> And thus alone with conscious meaning wear
>   The Indian's moccasin! (47)

In these final lines of the poem, Eastman suggests a desire, and even an ability, to literally and figuratively put the writer—and reader—in the shoes of Native Americans. The "careless" quality of the poem's joy suggests an irresponsible hope to escape civilization. This comes after first claiming the moccasin flower is off-limits, yet she is strangely unable to stop herself from completing the journey.

The trajectory of Eastman's poem is intriguingly parallel to that of her own life. When she wrote this poem as a girl living at a secluded farm in the Berkshires, a family friend was General Samuel Chapman Armstrong, who was among the founders of the Hampton Institute in 1868 (Sargent 17–18). Within five years, Eastman was teaching in the Indian department at the Hampton Institute. She soon moved to what is now western South Dakota to teach in day schools for Native American children—a system she felt was far superior to boarding schools (Sargent 28).[4] In this work, Eastman was remarkably progressive, a stance only reinforced by her becoming fluent in Lakota and routinely wearing moccasins, to the chagrin of other whites on the reservations (Sargent 30). Eastman, however, witnessed the aftermath of one of the greatest tragedies in Native American and US relations, the Wounded Knee Massacre, an event that surely ended any remaining "careless joy" she may have been feeling. Elaine assisted her future husband, Dr. Charles Eastman, as he administered to many of the casualties of the massacre. Charles Eastman, grandson of well-known portrait artist Seth Eastman, was a Dakota Indian who worked as a reformer and authored many books as well, including *Memories of an Indian Boyhood* (1902).

Elaine Eastman was instrumental in Charles's publishing career, as she edited and typed his manuscripts. Their arrangement is worthy of careful consideration in this book, as so many captivity narratives were produced in collaboration. After a long marriage that ended in a bitter divorce, Elaine explained the nature of their collaboration in a letter to her sister Rose: "He was and is the *author*—although he wrote very carelessly and would not even try to correct or revise, therefore I did *all* the drudgery [...] While I deliberately preserved his ideas and style, suppressing my own, still I should have been recognized as collaborator" (qtd. in Sargent 89). In total, Charles Eastman wrote nine books, only one of which listed Elaine as a co-author, and he produced no books after their divorce. Collaborative arrangements, especially in the nineteenth and early twentieth centuries, are often marked by a power imbalance that leads to an impression of single-authored texts. For all of Elaine Eastman's efforts to "put on the moccasin," her later outlook—she lived until 1953—regarding the bridging of cultural barriers was more pessimistic than that of her younger years, despite the six children she and Charles raised together.

Despite persisting worries about captivity on the frontier, by the end of the nineteenth century, most native orchids were called lady's slippers, rather than moccasin flowers. As historian Frederick Jackson Turner famously concluded in 1893, "[t]he stubborn American environment" and the "ever retreating frontier" had disappeared (59–60). The removal of Native Americans from lands that white settlers desired—especially those east of the Mississippi, which correspond with the primary natural range of these orchids—was all but complete. While the nineteenth century saw an often bloody resistance eventually give way to radically modified cultural and environmental landscapes, it is also true that Native Americans were and are still present across the US, as are several species of moccasin flowers, despite the relentless attempts to destroy, displace, and domesticate them.

Writing at the end of the nineteenth century, Mabel Osgood Wright, who also quotes Eastman, expressed regret over these changes to the land and its native inhabitants: "Ladies' Slipper is not a word in keeping with Hemlock and Beech woods, but the word Moccasin throws meaning into the black shadows and brings to mind the stone axe and flint arrow-heads found not long ago" (*Flowers and Ferns* 131). When Wright spies a moccasin flower along a roadside, she picks it, but only because of worries that someone else will see it and "follow the trail into the woods, and make the whole colony prisoners" (132). The language of captivity for Wright is used in opposition to her preferred setting for plants, which may be seen in the final word of her book's title: *Flowers and Ferns in their Haunts*. "Haunts," of course, is a word that was often invoked when describing the traditional camps of Native Americans.

Wright argues that the best course of action when finding a moccasin flower is to "look our fill, touch and caress them, then come away, telling their refuge only to the bees" (*The Friendship of Nature* 42).[5] For the moccasin flower is, to her, a "queen"[6]—not unlike the "Indian princess" label given to Sarah Winnemucca and other Native American women after Pocahontas—and "an orchid very shy of transposition, seldom living over the second season after its removal" (*The Friendship of Nature* 98).[7] A related comparison may be made with the historical figure of Pocahontas, who died less than a year after being brought to England as Lady Rebecca.

The mortality rate of the Cherokee on the Trail of Tears, or that of the Paiutes who were marched from Nevada to the Yakima reservation in the winter of 1872, may be seen as tragically exemplifying Wright's sentiments.

While Wright's knowledge of Native American culture is, by her own admission, superficial, she is insistent on the importance of local knowledge.[8] Wright argues to keep the name moccasin flower for the pouched orchids of North America: "when a common name, spicy with the odor of the new western world, is given to a plant, I think we should keep it [...] and call our little group of inflated pouched Orchids, Moccasin Flowers, instead of Ladies' Slippers, as Britton does, a general title which confuses their personalities with the European species" (*Flowers and Ferns* 131). Indeed, Wright's argument is similar to that of the many nineteenth-century writers who worked to forge a uniquely American cultural identity through emphasis upon America's environmental and cultural differences with Europe, especially Great Britain.[9] However, her argument is belated, as the common name of lady's slipper had by then all but supplanted the earlier usage of moccasin flower.

On the subject of naming, Susan Fenimore Cooper suggests that "if we wish that American poets should sing our native flowers as sweetly and simply as the daisy, and violets, and celandine have been sung from the time of Chaucer or Herrick, to that of Burns and Wordsworth, we must look to it that they have natural, pleasing names" (87). Cooper may have been thinking of Bryant, whom she admired enough to quote at length on the subject of migratory waterfowl (219–20). She also takes great pleasure in Native American place names (299–309). On a note that Wright would find problematic a few decades later—and that is diametrically opposed to the theme of *Flowers and Ferns in their Haunts*—Cooper relates the rapidly increasing scarcity of moccasin flowers, but also writes that on June 7th, "We were in good luck," because "we gathered no less than eighteen of the purple kind, the *Cyprepedium acaule* of botanists" (68). To the detriment of all native flowers, many nineteenth-century writers advocated such flower gathering, a habit especially destructive to such sensitive plants as these orchids, which do not send up flower stalks on an annual basis.

The one species that continues to be called a moccasin flower with any regularity, *Cypripedium acaule*, is also known as the pink lady's slipper. A difficult native orchid to transplant due to strict mycorrhizal specificity in the fungi of its roots, *Cypripedium acaule* is listed as endangered in Illinois, exploitably vulnerable in New York (where Cooper lived), and in Tennessee is listed as both endangered and commercially exploited. While its range is widespread, the plant is rare, and appears in only one US state with lands west of the Mississippi River, Minnesota (USDA).

The story of the moccasin flower provides a window on nineteenth-century debates over Indian removal policies. The supposed inevitability of the orchids' decline, addressed in part by attempts to domesticate them, parallels the often brutal policies that forced Native Americans off of their traditional lands. Despite these parallels, the reality was much more complicated than a poem like Bryant's "The Maiden's Sorrow" might suggest. The boundaries between white and Indian were not always rigid, and the debate over Indian removal played itself out in many areas of American culture.

My study of the cultural shorthand embedded in nineteenth-century discourse on moccasin flowers reaches a conclusion similar to that reached by Jane Tompkins regarding James Fenimore Cooper's novels: "the use of Indians to represent qualities that white America lacked [...] motivates the nostalgia for Indianness [...] even as they affirm the impossibility of union with the 'dusky' race and acquiesce in its extermination" (111). As seen in such works as *The Last of the Mohicans* and Eastman's poem, the desire to cross these boundaries was widespread. In the appendices to her book *The Ignoble Savage: American Literary Racism, 1790–1890*, Louise Barnett lists approximately ninety frontier romance novels published between 1793 and 1868 on the subject of Indian-white relations (197–203).

Over and over, literary works from the nineteenth century express fascination with Native Americans and with wilderness. While the use of the moccasin flower as a synecdoche for Native Americans suggests that the categories of nature and culture were explicitly linked, the possibility of moving beyond these highly stylized and symbolical terms and into a written expression of resistance to Indian removal and westward expansion can be found in nonfictional captivity narratives. Captivity narratives are among the most formative and popular texts in American literary history, and we can learn much about our own relationship to the natural world by studying how these influential texts worked to shape environmental attitudes and land use policies. Seen in the context of Indian removal and the changes in the North American landscape, many nineteenth-century captivity narratives worked to question the status quo through their authentic and sustained challenges to westward expansion. Because authors such as Mary Jemison, John Tanner, and Black Hawk told their stories from the authoritative position of a life on the frontier, their narratives were very difficult to reconcile with the conventions into which the moccasin flower more easily and silently fit. Environmental and literary writing about moccasin flowers led inexorably to issues of social justice; captivity narratives are similarly bound with issues of land use and the natural environment.

Notes

1. On the topic of Bryant's advocacy for conservation of the natural environment, see Michael P. Branch, "William Cullen Bryant: The Nature Poet as Environmental Journalist" (*American Transcendental Quarterly* 12 [Sept. 1998]: 179–97).
2. On Lucy Larcom as an environmental justice writer, see Karen L. Kilcup's chapter "Golden Hands: Weaving America" in her book *Fallen Forests: Emotion, Embodiment, and Ethics in American Women's Environmental Writing, 1781–1924* (Athens: U of Georgia P, 2013). Kilcup concludes that Larcom "uses her prose, and

especially, her poetry, to politicize nature, challenge social hierarchies, and lament the nation's accelerating industrialization and excess individualism" (196–97).

3. Many excellent examples of Indian boarding school narratives are available, and over a wide range of periods. One earlier narrative is Zitkala-Sa's "School Days of an Indian Girl," which appeared in *The Atlantic Monthly* (85:508 [Feb. 1900]: 185–94). A more recent example is Basil H. Johnston's *Indian School Days* (Norman: U of Oklahoma P, 1988).
4. Elaine Goodale Eastman's book *Sister to the Sioux: The Memoirs of Elaine Goodale Eastman 1885–1891* (Lincoln: University of Nebraska Press, 2004) recounts her years in South Dakota. Though posthumously published in a 1978 first edition, the book was primarily written during the 1930s.
5. "Telling the Bees"—literally informing your farm's beehive when a family member died—was a folk practice of the time. John Greenleaf Whittier's poem of the same name serves as an excellent description of this ritualized form of mourning.
6. The showy lady's slipper, a rare North American orchid and the state flower of Minnesota, is *Cypripedium reginae*, which affixes the Latin word for queen to this flower.
7. On the subject of Indian princesses, see Rayna Green, "The Pocahontas Perplex: The Image of Indian Women in American Culture" (*Works Massachusetts Review* 16 [Autumn 1975]: 698–714.)
8. In a 1905 lecture, Wright said, "It is to be deeply regretted that the tradition and folk-lore of the Indian tribes native to the region have not been better preserved. The Pequods were exterminated literally before they were known, and the nomenclature of our rivers, etc., is little understood" (qtd. in Philippon, "Introduction" 24).
9. See Ralph Waldo Emerson's "The American Scholar" or the Hudson River School landscape painting "Kindred Spirits" by Asher Durand, which depicts William Cullen Bryant with Thomas Cole communing in the wilds of North America.

1 Redemption deferred

American captivity narratives as environmental literature

The relationship between captivity narratives, ecocriticism, and environmental justice

Indian captivity narratives influenced environmental ethics from the very beginnings of written literature in North America, and extended beyond the debate over Native American removal. The captivity narrative was a remarkably popular genre that was very widely read both in colonial and antebellum America. More than two thousand extant nonfictional captivity narratives tell the story of people who lived in the wilderness as prisoners rather than going to nature because of a zeal for science, an aesthetic quest for the pastoral or sublime, a government mandate to explore, or a search for voluntary simplicity.[1]

In colonial times, especially, captivity narratives were written by whites upon their escape or ransom from Native American captors and described the captives' trials and tribulations before their eventual (often providential) restoration to civilization. By the nineteenth century, however, this generic formula was no longer uniform. An extraordinarily flexible genre, captivity narratives were often published in pairings that argued opposing sides of a given conflict, thus competing to fix meaning upon events.[2] Far from being consistently damning in their treatment of Native Americans and native landscapes, some captivity narratives resisted simplistic propaganda that reinforced westward expansion and attempted to justify the removal of Native Americans. Rather, these authors were motivated to publish by a desire to show the validity and importance of Native American lifeways and the corollary value of intact natural ecosystems. After opening with an examination of the dominant beliefs in antebellum white American culture regarding both Native Americans and the natural environment, the captivity narratives detailed later in this book are truly remarkable. I focus in these later chapters on a series of captivity narratives written before, concurrently, and immediately after President Andrew Jackson's Indian Relocation Act in 1830. Each narrative challenges the policy of relocation and stands as eloquent, enduring, and influential cross-cultural acts of resistance.

While difficulties are involved with any collaboration, captivity narratives were often able to bridge cultural, ethnic, and environmental barriers precisely because of their collaborative authorship. The captive subjects of my chapters, Mary Jemison, John Tanner, and Black Hawk, were all illiterate. Each chapter considers the unique textual history of a specific captivity narrative; this approach helps show the context-dependent goals of the authors. By looking at the reasons their stories were told on a national stage and by studying how and why the editors entered into a collaborative, cross-cultural agreement with the captive, we can better understand how these narratives were constructed and received. If their textual histories are sometimes murky, the environmental and ethnic milieu from which they emerged was no less so. In learning how these influential narratives function and evolve as environmental discourse, we also learn much about American culture and history. By focusing on the stories of people who self-identified as Native American, yet relied upon white editors for the publication of their narratives, I show how Indian relocation was textually resisted across ethnic and social boundaries. The rhetorical opportunities offered by captivity narratives allowed historically underrepresented groups such as women and ethnic minorities to influence policies of Indian relocation and reform. All of my chapters focus on individual narratives that were republished in the 1990s—and studied by critics—as American Indian autobiographies, further complicating the already complex issues of genre and collaboration that characterize so many captivity narratives. My conclusion looks at how instances of captivity have also had long-reaching influences on communities of people, and how environmental restoration can influence memorialization of historic injustices.

Elaine Goodale Eastman's poem "Moccasin Flower," discussed in the prologue, points to a longstanding reason why captivity narratives remained so popular for so long in American history. Readers were fascinated by the idea of, in the words of Philip Deloria, "playing Indian." Deloria suggests that the American colonists who dressed as Indians and rioted in protest of oppressive laws were "conflating Indians and land" and that their disguises allowed them to "cross the boundaries of law and civilization" (26). While Deloria points out that crossing these boundaries simultaneously inscribes them, one fear early Americans voiced was that captives would choose against returning to white culture once ransomed. For example, J. Hector St. John de Crèvecoeur points out, through his narrator of James, the American Farmer, that "thousands of Europeans are Indians, and we have no examples of even one of those aborigines having from choice become Europeans" (214). When he is forced to abandon his farm due to war, James's only hesitancy regarding a move westward and living with Native Americans is that "the imperceptible charm of Indian education may seize my younger children and [...] may preclude their returning to the manners and customs of their parents" (219). As Gary Ebersole posits in a chapter called "Going Native, Going Primitive," the striking pattern that Crèvecoeur's James worries over posed a serious challenge to white Americans, as this pattern "ran counter to the assumption that theirs was an undeniably superior (and infinitely desirable) civilization" (191).[3] The most comprehensive study of Indian captivity points to a related and even more present concern for white Americans: "Conservative estimates place the number of captives taken by Indians [during the colonial period and the nineteenth century] in the tens of thousands" (Derounian-Stodola, Levernier 2). Not only were white Americans regularly taken captive by Native Americans, whites had the additional fear that their children, once taken captive, might choose to remain within the Native communities to which they became acculturated.

The frequency of captivity, spurred by the very real and significant resistance of Native Americans to white encroachment, is often forgotten today, as it does not fit the heroic rhetoric employed by many twentieth-century depictions of "how the West was won." When taking a closer look at the very real danger of captivity on the frontier—and the anxiety created by the existence of "white Indians"—we can better understand why captivity narratives were so widely read. The term "white Indian," simply stated, was widely used in the nineteenth century to describe white captives who became acculturated into a Native culture, many of whom chose to remain with adoptive Native American families. These people were a source of fascination to white society, as seen by the popularity of captivity narratives of white Indians such as Mary Jemison, John Dunn Hunter, and John Tanner. Fictional stories of such people were also common during this period; one melodramatic short story published in 1827 was simply titled "The White Indian."[4]

Captivity narratives depict landscapes from a perspective uninfluenced by choice: for these writers, nature is a place they are forced to inhabit. Perhaps this is why ecocritics have rarely engaged themes of captivity. Yet, there is much in common between captivity narratives and the literary nonfiction that much early ecocriticism has privileged: from a willingness to list information in appendices to a first-person autobiographical narrator, these genres are suited to similar types of analyses. The richness of detail in many captivity narratives rivals some natural history texts. Lawrence Buell does point out in *The Environmental Imagination* that "nature has historically been not only directly exploited but also the sign under which women and nonwhites have been grouped in the process of themselves being exploited" (21), and he observes that Mary Rowlandson is a potential counter to the "notion of androcentric pastoral escape as the great tradition within American literary naturalism" (25). Rowlandson's *True History of the Captivity and Restoration of Mrs. Mary Rowlandson* (1682) is generally considered to have helped create and define the captivity narrative genre, casting a long shadow of influence over later American literature. This title is what is most often used in contemporary reprints and is based on the fourth edition of Rowlandson's narrative, published in London in 1682. My quotations are from the second edition, published in Cambridge, Massachusetts, also in 1682.

In the rest of his seminal book, Buell examines Thoreau's *Walden* and other texts that emphasize tropes of escape or withdrawal from civilization into nature. This notion of "escape," of course, is in direct contradiction to a state of captivity in a wilderness landscape. As critics have recognized in a related genre, the slave narrative, "nature represented a potential refuge and means of self-affirmation *and* a place of dehumanization and danger" (Kilcup 78; emphasis in original). Buell's appendix, "Nature's Genres: Environmental Nonfiction at the Time of Thoreau's Emergence," does not mention captivity narratives—which were much more popular and established than were any of the genres he does mention, such as literary almanacs or homiletic naturism. Therefore, my inversion of a standard trope of environmental literature—pastoral escape—can be seen as providing an alternative tradition within the environmental literatures of early America.[5] Rather than describing nature as an idyllic space within which to reflect in solitude, many captivity narratives regularly describe the great personal hardships their authors endure in the wilderness while being integrated into an unfamiliar human community. This does not mean that the wilderness is not described; rather, these authors describe their personal experience of nature using a different set of parameters. Instead of celebratory passages on the beauty of wilderness, captives write about the uses of various plants or animals for clothing, shelter, and food. Wilderness is not described as a place apart from humans, but rather as a viable home. Despite the frequent suffering of captives, they indicate with authority that wilderness is a place that can sustain culture and inhabitation. Many captivity narratives featured additional materials, often in the form of appendices, which detailed ethnographic, linguistic, and scientific information. Tanner's narrative, for example, featured nearly 150 pages of such materials.

This underlying message of wilderness as home is at times undermined by colonial responses of captives to turn inward and hold onto familiar intellectual and spiritual patterns of thought in order to maintain their identity. Rowlandson meditates upon her Puritan identity so intensely that she completely forgets her surroundings:

> I cannot but remember how many times sitting in their *Wigwams*, and musing on things past, I should suddenly leap up and run out, as if I had been at home, forgetting where I was [...] But, when I was without, and saw nothing but *Wilderness*, and *Woods*, and a company of barbarous heathens; my mind quickly returned to me (88; emphasis in original).

Rowlandson struggles to maintain her former identity, a struggle she ultimately and perhaps inevitably loses. After being successfully ransomed, she has a parallel experience, writing that "*I used to sleep quietly without workings in my thoughts, whole nights together, but now it is other wayes with me.* [...] I remember in the night season, how the other day I was in the midst of thousands of enemies" (111; emphasis in original). Rowlandson thus dwells upon her captivity even after redemption, a trope of many captivity narratives.

Despite Rowlandson's struggles and what is arguably her high degree of transculturation, I find the language used in the final paragraphs of her narrative not so different from Thoreau's conclusion in *Walden*: Rowlandson struggles to make sense of her life and her private thoughts, and Thoreau argues that "it is easier to sail many thousand miles ... than it is to explore the private sea, the Atlantic and Pacific Ocean of one's being alone" (300–01). Thoreau's wine is that of a "purer wine, of a more glorious vintage, which they had not got, and could not buy" (310), while Rowlandson experiences "the Wine of astonishment" (112). Both writers are intoxicated in some way by experiences that are intensely personal and derived from considerable time spent in the woods. The difference, of course, in these two extraordinarily influential American writers, is that Thoreau's writing describes a life he chose for himself, while Rowlandson was denied agency by her captors and, at least to some extent, by her own Puritan community.

As seen in the wording of Rowlandson's title, a crucial trope of colonial captivity narratives is the restoration of the captive to civilization: through the physical and spiritual restoration of the captive to his or her colonial family and minister, the cultural values that a colonial audience holds are affirmed. The nineteenth-century writers and editors whose work I examine most closely in this project use this earlier convention to great effect, often thwarting the reader's expectations by either failing in their attempts at restoration or refusing even to attempt such a formulaic conclusion. John Tanner, for example, was plagued by illness and language barriers throughout a brief restoration in Kentucky with his biological family before returning to Indian life in the Michigan Territory, the place upon which the final chapter of his book is centered. Other narratives, such as that of Mary Jemison, have tangled textual histories suggesting that the affirming closure of restoration remained irresistible to a popular audience, even in nineteenth-century America. While Jemison successfully appropriated the genre of the captivity narrative to preserve her land rights, the irresistibility of a formulaic conversion story is evidenced by later, expanded versions of the narrative produced after both the original editor and Jemison died. In these later versions, Jemison reportedly returned to Christianity just before her death. Although this restoration is dubious and impossible to confirm, it was an episode that was repeatedly expanded and emphasized by a series of later editions that argued for the primacy of whiteness to Jemison's identity (Namias, *White Captives* 157, 169).

The rewriting of many captivity narratives—often by male editors of female-authored texts—also helps explain the relative scarcity of ecocritical attention. Much early analysis of captivity stories concludes that "our national preoccupation with female rescue [is] a mere cover story, a pretext employed to justify the sanguinary pleasure our pioneers took in the slaughter of the continent's natives and the decimation of the wilderness. That is: first we conquered, and then we made up a fiction of defiled womanhood to rationalize it" (Faludi 212–13). Thus, captivity narratives have been treated as complicit in the removal of Native Americans and the destruction of native ecosystems. However, this analysis has been found inadequate because many captivity narratives suggest that women had much more agency than such an assessment allows.

In fact, some captives find their identity changed so radically that they come to view wilderness, rather than civilization, as their home. With this outward shift in perspective comes a corresponding inward shift. Captivity narratives outline the possibility—and appeal, for sympathetic readers—of changing one's identity from white to Native American. Thus, I argue that captivity narratives bridge an identity divide between white audiences and native conceptions of the natural world and of ethnicity. On the one hand, Native Americans are seen as part and parcel of Nature; on the other, native peoples are reaching out to white American audiences through captivity narratives in an active attempt to materially change white cultural and environmental norms. This attempt is especially clear in those texts that feature white Indians who refuse to reintegrate into white society. These narratives materially destroy divisions between race in the surprisingly numerous accounts of captives who refuse to be redeemed.

Ecocritics have only hinted at the environmental implications of captivity narratives. While authorial intentions vary considerably from text to text, the earliest colonial narratives owe much of their form and rhetoric to Rowlandson. Rowlandson's narrative, originally titled *The Sovereignty and Goodness of God, Together with the Faithfulness of His Promises Displayed*, can be described as the first bestseller published in America. Rowlandson's relationship with the natural world—which she repeatedly calls a "howling wilderness"—shifts over the course of the narrative. A close reading suggests that she becomes increasingly able to discern what may be edible in nature, for example—or is at least increasingly flexible in her tastes, as I will show shortly. This shift is by no means unique to her narrative, as the interpretation of the American landscape is integral to the captivity narrative as a genre.

Susan Howe responds to Rowlandson's narrative with mild frustration:

> We will read no lovely pictures of the virgin forest; no night fishing, no deer hunting, no wildlife identification, no sunsets, no clouds of pigeons flying. ... It is always either snowing or raining, muddy and dreary. Landscape will never transfix her. The beautiful Connecticut River is just another barrier to get across. Rowlandson's apprehension of nature is an endless ambiguous enclosure. (96)

Howe searches for the notes of creative wildness in Rowlandson that she finds more readily in the expressive verse of Emily Dickinson (and, based upon the allusions in the quotation, also found in James Fenimore Cooper's *The Pioneers*), concluding that Rowlandson's words stand as a "bloody fragment of the world ... a relentless origin" (96). The aspects of Rowlandson's text that are seen as creating a new genre and tradition in American literature are what most critics focus on, rather than on its environmental aspects.

Ecocritics have long privileged writers who provide realistic and detailed observations of nature. Thus, the Biblically inflected landscape of Rowlandson's narrative allows critics to avoid the important point that Indian captives confront a wilderness that is not devoid of humans, but rather has had continuous human inhabitation for thousands of years. Rowlandson often focuses on food, a common trope of captivity narratives, as hunger to the point of starvation is common during war and captivity. She describes the diet of her captors, and increasingly, herself.

> They would eat Horses guts, and ears, and all sorts of wild Birds which they could catch: also Bear, Venison, Beaver, Tortois, Frogs [...] yea, the very Bark of Trees [...] I can but stand in admiration to see the wonderful power of God, in providing for such a vast number of our Enemies in the *Wilderness*, where there was nothing to be seen, but from hand to mouth. (106; emphasis in original)

David Mazel discusses Rowlandson's narrative in *American Literary Environmentalism*. Mazel is concerned with the origins of American literary and social conceptions of wilderness and sees Rowlandson as writing the land out of existence. Rowlandson's wilderness is not a landscape, but rather "a *condition* of loss and bewilderment" (51; emphasis in original). He writes that "the wilderness is never the sort of landscape that subsequent travelers might recognize by her descriptions of it; it is precisely 'the wilderness where there was nothing to be seen'" (Mazel 48). That last phrase, "the wilderness where there was nothing to be seen" is from the extended Rowlandson quotation given previously, but Mazel skips over the context, which suggests instead that many things—including food—are visible to an experienced eye.[6] Mazel does see change occur over the course of the narrative. He finds that after Rowlandson receives a copy of the Bible, the landscape suddenly springs into focus and is read as an analogue to a Biblical landscape (51). This correspondence, conditioned by the Puritan system of typology, allows Rowlandson's text to function as a proto-environmental narrative with openings and ambiguities that critics have only recently begun to recognize. For example, Mazel acknowledges that Rowlandson ultimately "preserves her experience of wilderness" through her preference for the immensity of her trials rather than a bland desire for a less challenging experience (58). Or, in Rowlandson's own words, "the Wine of astonishment: like a sweeping rain that leaveth no food, did the Lord prepare to be my portion" (106). Rowlandson's diction thus offers a metaphorical language for the wilderness that is no longer typified by "lack," which constitutes a subtle but important shift from the opening pages of her narrative.

Both Howe and Mazel are further disappointed by Rowlandson's refusal to engage in what Howe calls "wildlife identification." Mazel suggests that "subsequent travelers" will not be able to use her text as they explore the same forests in which she was a captive. The search for a literary counterpart to actual places suggests important touchstones for environmental criticism. In this sense, first-wave ecocriticism, as exemplified by Buell's *The Environmental Imagination*, echoes the modernist manifesto of William Carlos Williams: "No ideas but in things." In part because of their concern for close observation of the natural world, ecocritics have rarely considered captivity narratives as environmental literature. Following Rowlandson's creation of the Indian captivity narrative, few narratives by colonial captives satisfy ecocritical demands for identification and recognition of real places, flora, and fauna.

One interesting counterexample was advanced by Alexander Starre, who notes that "the genre has not appeared attractive enough for environmentally interested scholars" ("Wilderness Woes" 290). He notes, somewhat against my own argument, that Rowlandson does not accept the wilderness as a place that people live in, despite her ability to survive. Starre admits that the "clear border between wilderness and civilization" that Rowlandson maintains is, in fact, prone to "paradoxical negotiations" ("Wilderness Woes" 282). The captivity narrative Starre suggests deserves more attention for ecocritics is John Gyles's *Memoirs of Odd Adventures, Strange Deliverances, Etc.* Published in 1736, Gyles was a captive from 1689 to 1698 in what is now Maine. Rather than solely serving as a text on wartime deprivation, Gyles devotes a chapter to natural history, calling it "A Description of Several Creatures commonly taken by the Indians on St. John's River" (24). Starre concludes that Gyles's text "could not effectively counter the anti-nature rhetoric of other narratives like Rowlandson's since it failed to reach a large readership" ("(Forced) Walks on the Wild Side" 30). Elsewhere, Matthew Wynn Sivils examines the conflicting attitudes that Starre takes on by looking a series of fictional captivity narratives that he calls "proto-environmental" (39). Sivils asks how the Indian captivity narrative can "become complicit in the initiation of European Americans in what was to them a land at once beautiful and hostile?" (41).

Such analysis, however, is the exception rather than the rule. I believe the kind of "positive argument" feminist critics have found for reading women's captivity narratives has been lacking in ecocritical scholarship on the same texts. The mere existence and popularity of these texts means that their relevance to environmental concerns in early American literature is significant. Captives were given a voice, which makes possible the study of popular attitudes about the natural world from the perspective of women, the poor, and ethnic minorities. I argue that the study of captivity narratives—even those a "subsequent traveler" would be unable to use as a field guide—reveals a great deal about the environmental attitudes engendered by the text. In these texts, we glimpse the lifeways of migrant and nomadic people, including food cultures and subtle land management practices. The affection for and deep knowledge of specific places possessed by individual Native

Americans can be seen throughout. We learn about the disastrous wartime destruction of food caches and fields relied upon and cultivated by Native Americans. Alternative attitudes toward the natural world are often on display; animals such as bears and wolves, for example, are at times accorded a level of respect and even deference not found in exploration narratives or natural history texts.

Although Rowlandson's narrative set many of the genre's conventions into motion, Álvar Núñez Cabeza de Vaca's narrative of the disastrous 1527 Narvaez expedition is arguably the earliest autobiographical captivity narrative. Cabeza de Vaca depicts many places and things that have been identified by modern scholars with their counterparts in the Gulf coast and what is now Texas and Mexico. He famously records the first European account of the American bison, for example. Scholars of captivity narratives read Cabeza de Vaca for the complexity of his interactions with a multitude of Native American cultures. Indeed, after learning six distinct native languages, Cabeza de Vaca continues to encounter people with whom he cannot communicate. The ability to differentiate between Native cultures and individual humans—something for which captivity narratives are often noted and that is in opposition to prevailing Euro-American attitudes—can be fruitfully compared to the ability of Cabeza de Vaca to differentiate between distinct species of deer, pinyon pines, and even mosquitoes. Upon nearing Spanish outposts at the end of his nine-year odyssey, Cabeza de Vaca displays both transculturation and attunement to the environment:

> We traveled through much land and we found all of it deserted, because the inhabitants of it went fleeing through the sierras without daring to keep houses or work the land for fear of the Christians. It was a thing that gave us great sorrow, seeing the land very fertile and very beautiful and very full of waterways and rivers, and seeing the places deserted and burned and the people so emaciated and sick, all of them having fled and in hiding. (156)

Cabeza de Vaca here makes an extraordinarily early and perceptive connection between wilderness and the human rights of native peoples. While such connections are clearly stated in many nineteenth-century captivity narratives, they are more often implicit in colonial narratives. In this respect, Cabeza de Vaca's narrative may be seen as more farsighted than Rowlandson's.

Rachel Plummer, captured at Parker's Fort in Texas by Comanches in 1836, wrote in great detail about her nearly two years in captivity. Plummer's cousin, Cynthia Ann Parker, was also captured in this raid. Parker would remain with the Comanche until forcibly taken by Texas Rangers in 1860; after numerous thwarted attempts to return home to her Comanche family, she starved herself to death in 1870. Her son Quanah Parker went on to become a leader of the Comanche and in 1910 lobbied successfully for his mother's remains to be moved to the Comanche cemetery near Fort Sill, Oklahoma. Her uncle James Parker's self-promoting search for her was later a primary inspiration for the novel and film *The Searchers*, and the main character Ethan Edwards, as portrayed in the film by John Wayne (Faludi 200–08).[7]

Plummer's narrative, which swings wildly from melodramatic scenes of horror to careful natural observations, contains many passages offering rich and realistic descriptions of nature:

> Notwithstanding my sufferings, I could not but admire the country [...] a range of timber-land from the waters of Arkansas, bearing a southwest direction, crossing the False Ouachita, Red River, the heads of Sabine, Angelina [...] &c., going on south-west, quite to the Rio Grande. This range of timber is [...] 5–35 miles wide, and is a very diversified country; abounding with small prairies, skirted with timber of various kinds—oak, of every description, ash, elm, hickory, walnut, and mulberry. (338–39)

Upon concluding this survey, Plummer immediately asks the reader to "bear with me in rehearsing the continued barbarous treatment of the Indians" (339). She tells of her child being tortured and then being taken away mysteriously, implying that the child has been murdered. Shifting between her personal sufferings and her surroundings, Plummer sees the natural world as a bridge that allows for healing. She pivots from flat recitations of trauma to enthusiastic, detailed observations of nature. As Plummer becomes more immersed in her surroundings, she seems able to find pleasure in the quality of her new experiences. Other, especially original, passages include listings of fifteen large mammals that Plummer finds interesting and worthy of description, including bears, buffalo, antelope, and wolves. Her precision is such that she describes four types of bears and five of wolves (she is listing them based on coloration and habits, not species differentiation). As the narrative progresses, the "rehearsals" of Comanche barbarity fall away, yielding to pages of uninterrupted descriptions of the landscape and the peoples and animals she contacts. Not unlike Rowlandson's narrative, complaints and mourning appear with less frequency as the text progresses, mirroring the captive's possible transculturation and level of comfort with their new life within Native American culture.

Plummer also tells a lengthy story of spelunking. The adventure she has in a cave in present-day Colorado is quite remarkable.

> Reader, you may fancy yourself viewing, at once, an entirely new planetary system, a thousand times more sublime and more beautiful than our own, and you fall far short of the reality I here witnessed. I soon discovered that these lights proceeded from the reflections of the light of the candle by the almost innumerable chrystalized formations in the rocks above. (349)

Plummer enters the cave only after beating back a Comanche woman who tries to stop her from entering it. Plummer travels several miles into the cavern; halted by a subterranean waterfall, she stops to rest. Upon waking, she climbs to the surface to discover she had been in the cave for more than two days. She even claims that

> now in the city of Houston, surrounded by friends and all the comforts of life, to sit alone, and in memory, retrace my steps in this cave, gives me more pleasurable feelings than all the gaudy show and pleasing gaity with which I am surrounded. The impressions made upon my mind in this cave, have since served as a healing balm to my wounded soul (351).

Plummer explicitly identifies nature as a source of strength and solace, thus echoing both Rowlandson and Thoreau.

In captivity, Plummer turns to nature to distract her from the treatment of the Comanches; after being ransomed, she continues to find nature more soothing and attractive than the "gaudy show" of her city and friends. Indeed, she even saved a few thorns that she collected as mementos of her time in the wilderness.

> I have two of them now that I have caught many a fish with. I took them off the bush myself ... I have often been offered five dollars for one of them, but I have never been induced to part with them. They often bring to my recollection the distant country where I obtained them (358).

Plummer's tone is often similar to that of an enterprising traveler in search of a sublime view, rather than that of a desperate captive; only in her most stale and contrived passages does she rehearse the conventions of a melodramatic captivity.

Plummer does seem to resist transculturation, including fighting her captors violently and then defending herself in a sort of Comanche legal trial regarding the fight, thus circumventing both the stereotype of the damsel in distress and the fears of readers that she will be assimilated. The prevalence of such episodes has prompted many feminist critics to reclaim captivity narratives as texts that show remarkably strong, self-reliant female authors. Puritan captive Hannah Dustan, for one, is so powerful that she kills and scalps ten captors and successfully demands a bounty for the scalps from the Puritan government upon her return. Dustan's narrative was published in 1697 by Cotton Mather. As Derounian-Stodola and Levernier dryly explain before describing Dustan's story, "[t]he captivity literature presents various strategies by which women outwitted their (male) captors" (133).

Plummer struggles to put into English the experiences of nature and landscapes that she understood through a different cultural lens, writing, "I wish I had language to give a fair description of this part of the country, with its present inhabitants" (358). Plummer even argues obliquely against white settlement: "The healthiest looking Indian I ever saw, lived here. Notwithstanding it is a healthy country, I do not think it will ever be settled by white men, as I saw nothing to induce white men to settle here" (358–59).[8] While she labors to create an objective account of her experiences, Plummer ultimately admits that she cannot fully describe what she saw to white readers.

At this point, however, my argument diverges from the few existing critical assessments of how the environment functions in captivity narratives. I believe that critics who struggle to find a realistic representation of wilderness in texts such as Rowlandson's narrative miss the important point that captives are forced to confront the reality of a North American wilderness that is not devoid of humans, but rather has been continuously inhabited for many thousands of years. Captivity narratives often were absorbed into the popular imagination of white Americans with the assumption that wilderness and Native Americans could be conflated as a monolithic whole. Clearing the landscape for white settlers thus meant a clearing of Native American inhabitants. Examining captivity narratives that contain cross-cultural resistance to Indian relocation exposes links between environmental and social justice.

While environmental justice scholarship most often examines contemporary texts, one of its most influential proponents, Joni Adamson, agrees with the scope of the connections I make:

> [T]he origin of what has come to be known as the environmental justice movement is much older than usually acknowledged. It could be said that this movement began well over five hundred years ago, when indigenous peoples first began resisting exploitation and fighting for resources and lands that were being stolen by European colonizers (196).

Captivity narratives almost always appear in the context of contact followed closely by resistance, as territorial conflicts usually form the backdrop of their textual production. One major reason for taking captives was to replace family members killed by war or epidemics associated with the Columbian exchange (Derounian-Stodola, Levernier 4). Despite Adamson's willingness to extend environmental justice concerns to earlier periods, she does not discuss early texts in detail. Furthermore, environmental critics, including those who consider nineteenth-century literature, have focused primarily upon literary nonfiction by white writers. Captivity narratives thus provide important opportunities to ecocritics. They offer a rare cross-cultural textual space in which land ownership and use was actively contested. They serve as a testament to the resistance of Native Americans and sympathetic Euro-Americans to the supposed inevitability of Indian removal; as such, these nonfictional accounts also represent an important environmental legacy.

One of the most important moments in the contemporary field of environmental justice was the publication of the 1987 report of the United Church of Christ Commission on Racial Justice. This report showed that over 50% of all ethnic minorities in the US live in an area with an uncontrolled toxic waste site (Adamson et al 4). Minorities continue to be correlated with toxic lands today, and the relationship continues to be one of marginalization through the process of environmental racism. In important ways, the segregation of minorities into these toxic landscapes can be traced to nineteenth-century captivity narratives, which are entangled within larger debates about ownership of land. Captivity narratives also document the relocation of Native Americans into urban areas as well as reservations which are located in the most remote parts of the continent—the very spaces that today are often sites of pollution and industry.

Rather than remaining on the periphery of land debates, captivity narratives were widely read and also highly influential in conditioning American attitudes toward the frontier. Gordon Sayre writes that "to satisfy its strong appetite for firsthand representations of Indian life, the nineteenth-century American audience was fed a steady diet of captivity narratives by whites" ("Abridging" 483). Henry Rowe Schoolcraft, in his 1848 book *The Indian in His Wigwam*, describes the major influence that captivity narratives had on his imagination as a child in Vermont:

> My earliest impressions of the Indian race, were drawn from the fireside rehearsals of incidents which had happened during the perilous times of the American revolution … all inseparably connected with the fearful ideas of the Indian yell, the tomahawk, the scalping knife, and the firebrand. […] These early ideas were sustained by printed narratives of captivity and hair-breadth escapes … so that long before I was ten years old, I had a most definite and terrific idea impressed upon my imagination of what was sometimes called in the native precincts, "the bow and arrow race." (64)

Once on the frontier, Schoolcraft admits, his impressions changed and diversified rapidly. While many early American captivity narratives functioned as propaganda in favor of westward expansion, often connected (as in Schoolcraft's recollection) with the creation of the republic, others made passionate arguments for social and environmental justice. This alternative tradition grew more and more apparent in the nineteenth century as the stylized conventions were discarded or used as an opening to describe personal experiences.

Captivity narratives became doubly significant as their authors were forced to engage with the reality of their status as hostages of both a different cultural group and of the land. Captivity narratives can break down the dichotomy between human and nonhuman elements, and, in the cases of captives who refused to be redeemed, the divisions between races. Rather than limiting myself to a study of those descriptive passages from various captivity narratives that appear most ripe for environmental criticism because of their close attention to the natural world, I instead examine issues of environmental justice throughout the texts.

Thus far, scholarship on captivity narratives has focused on achieving a fuller understanding of ethnic and gender relations in American culture; my project argues that captivity narratives have similarly affected land use policy and environmental attitudes. Jeffrey Myers writes that

> in order for environmental literature to become a foundation for change, writers, critics, and activists must greatly expand what we mean by 'environmental issues' to include the ways in which racism and environmental destruction have been linked under a larger pattern of Euroamerican domination for five hundred years (140).

Captivity narratives illuminate Myers's link between racism and environmental destruction by depicting a variety of attitudes about the frontier—both the land and its inhabitants—over time and in dialogue with historical events. Not only do captivity narratives describe the early American landscape in unique ways that differ from accounts offered by such natural history writers as William Bartram or Henry David Thoreau, they also offer perspectives from women and ethnic minorities who rarely had opportunities for publishing on the environment outside popular genres such as the captivity narrative. While many captivity narratives were written to justify the usurpation of lands traditionally used by Native Americans and to reinforce cultural assumptions about white superiority, the narratives I examine offer an alternative argument.

Captivity narratives have helped create a legacy for specific historical sites that were preserved in part because of the popularity of the texts themselves. In other words, the memorialization and preservation of specific areas was made possible because captivity narratives raised awareness about the traditional use and importance of these places. Some of the places described in the texts are also under Native American management today, either as part of a reservation or under partnerships with federal and state agencies. In *Conserving Words: How American Nature Writers Shaped the Environmental Movement*, Daniel J. Philippon asserts that nonfiction writers (he spends a chapter each on John Muir, Aldo Leopold, Edward Abbey, Theodore Roosevelt, and Mabel Osgood Wright) made a difference in how American environmentalism developed: "Not only were these writers representing their own human ideas of nature … the words they employed also had observable effects; they achieved results and mattered in practice. They were *conserving words*" (14; emphasis in original). I make a similar argument about many captivity narratives. While not all captivity narratives are explicit arguments for conservation, they do share in a cultural project of explaining and shaping "human ideas of nature" (Philippon 14).

As captivity narratives are also essentially nonfictional works, it is possible to similarly trace the conservation legacy of these texts.

The authors and editors of the captivity narratives I examine did often view their stories as a plea for environmental and social justice, though their lives sometimes ended before they could see the importance their words would have in preserving land and native lifeways for later generations. The legacy of these texts is not tragedy, but renewal. These collaborative texts, which resist the injustices of westward expansion and a hierarchical view of race, found an audience hungry for alternative models of a more inclusive and less exploitative American society. This audience remained substantial, growing in size and influence over the decades following publication of the first editions of the texts. I read the continued popularity of these texts—and the legacy of conservation and cross-cultural connections they have engendered—as the promise of environmental and social justice that has been slow in fulfillment. As these texts testify, the struggle for justice was never ceded. Through the resistance offered by each captivity narrative, the groundwork was laid for ethnic and environmental renewal.

American literature, the Indian relocation act, and collaboration

The texts upon which I focus were originally published during the 1820s and 1830s. These decades—characterized by crucial national debates regarding Indian removal and how the rapidly growing nation should expand—have been somewhat underrepresented in literary criticism. Arnold Krupat's 2009 book chapter "Representing Indians in American Literature, 1820–1870" explores how white writers failed to forthrightly engage the issue of Native American removal. "What I find extraordinary," writes Krupat, "is the degree to which, when it comes to the actual, detailed, year by year history of Native American removal and dispossession in the nineteenth century, both the canonical and the lesser authors of the period, have extraordinarily little to say" (*Ethnocriticism* 51). Krupat considers several major authors: "[T]hey very rarely engage with [Native Americans] as contemporary historical, cultural and social beings—and when they do, as we have noted even with Melville, Thoreau, and Fuller, it is still to see them as living anachronisms or last remnants" (52).[9] This charge is accurate, although somewhat simplified. While we study Emerson's *Nature*, we should also remember that Emerson wrote an open letter to President Martin Van Buren in 1838 entitled "A Protest against the Removal of the Cherokee Indians from the State of Georgia."

Krupat's subject of literary representations of Indian removal is not an entirely new topic, though his insights are finely developed. Lucy Maddox's 1991 book *Removals: Nineteenth-Century American Literature and the Politics of Indian Affairs* is an earlier and more exploratory treatment. In her introduction, Maddox struggles to come to terms with the divide she finds between traditional academic disciplines and alternative perspectives, noting that "our working definition of American literature has not yet been able to accommodate Indian texts, oral or written, very comfortably" (3). Maddox reaches many of the same conclusions as does Krupat: the major American authors of the nineteenth century see Native Americans as doomed to a fate of "civilization or extinction." Maddox makes remarkably detailed arguments that authors such as Melville were drawing heavily on debates about Indian removal in texts such as "Bartleby, the Scrivener." She connects the unwilling Bartleby to the Cherokee through a textual comparison of the historical figure of the first Commissioner of Indian Affairs, Thomas McKenney, and the unnamed narrator of Bartleby. Melville, Maddox thus argues, is actually engaged with the debates over Indian removal:

> As a result of the landlord's ultimatum, Bartleby is taken away on his own private Trail of Tears; his is conducted in 'a silent procession' (42) across what the narrator has already referred to, significantly, as 'the Mississippi of Broadway' (28) and led to the place where he will be confined, and where he will (as many of the transported Cherokees did) starve to death. (73)

Her argument is made persuasive by the thoroughness of her research regarding McKenney's 1846 *Memoirs*, and the possible allusions Melville may have made to this text (and Washington Irving's *Astoria*) in "Bartleby," published in 1853. Yet Maddox avoids confronting the tension with which she opens. This same tension comes from a dynamic Krupat so clearly identifies: very few texts by Native Americans in the early nineteenth century are studied by literary critics, and few white Americans of the period wrote explicitly about Indian removal. If "Bartleby" was inspired in the ways Maddox suggests, Melville was still primarily influenced by the written efforts of McKenney, who was writing not to resist Indian removal, but rather to clear his name after he was terminated in 1830 by President Jackson. The texts upon which I focus address these concerns much more fully and directly as they are the products of cross-cultural collaborations that explicitly address issues of Indian removal based on first-hand experience.

My chapters closely examine texts produced during the nine-year period from 1824 to 1833. By 1824, a large readership of middle-class whites was rapidly consuming texts that discussed Native American life, and was sufficiently removed from the frontier to no longer feel threatened by captivity or war. Yet Indian relocation was a major debate waged both locally and nationally. While the 1830 Indian Relocation Act is an important marker of acts of relocation that occurred throughout the continent, captivity narratives are an artifact of popular culture that show the choices and sacrifices made in the course of individual lives. The literary individualization of the relocation project affected a wide readership, as evidenced by sympathetic reviews in periodicals of the day such as the *American Quarterly Review* that do not speak out against Indian relocation in any sustained manner. In other words, by the 1820s, captivity narratives that resisted relocation were well-received and seen as significant contributions to a competitive literary marketplace, which developed alongside the political debates.

Even when called inevitable, Indian relocation remained a highly contentious process. Drawing attention to the notion of God's providence and the long-standing belief by many white Americans that it was their destiny to expand westward, Krupat cites the full title of the Indian Relocation Act: "Act to Provide for an Exchange of Lands with the Indians Residing in any of the States or Territories, and for their Removal West of the River Mississippi." Krupat's close reading of the title examines the common etymological root of the words "provide" and "providence" (*Ethnocriticism* 136–37). Captivity narratives have long emphasized the idea of providence as a form of protection or salvation for captives in distress. The popularity of captivity narratives in the 1820s and 1830s, coupled with the importance of their subject, suggests a deliberateness to using this genre to engage with the debates surrounding the drafting and passage of the Removal Act. The passage of the bill was more uncertain and contentious than is commonly believed: the final vote totals were 28–20 in the Senate and 103–97 in the House (Krupat, *Ethnocriticism* 143).

My argument that many captivity narratives work to individualize Indian removal differs from early work on captivity, such as that by Richard Slotkin, which argues instead for captivity as a mythmaking process (63–64). Admittedly, many of the narratives I explore are somewhat unusual, and so wide a range of captivity narratives is more difficult to find before the nineteenth century. Much of the focus of earlier critics has been on colonial narratives, rather than nineteenth-century texts. Interestingly, the qualities of resistance to the status quo in the narratives I examine increased interest in them, rather than relegating them to obscurity. Because of the authenticity of an individual's captivity, the rhetorical space needed to question assumptions regarding gender and cultural norms was more readily granted.

In addition to engaging issues of environmental and social justice, my project also studies nineteenth-century captivity narratives for important textual reasons. Captivity narratives often were the product of editorial/authorial collaborations that, in some instances, exemplify tolerance, cooperation, and sustainable cross-cultural/trans-cultural lifestyles. Lisa Logan's definition of cross-cultural collaboration is useful here. In texts produced by cross-cultural collaboration, she writes, "the narrative self suspends accommodation or explanation of the other and allows difference simply to exist in the text" (473). Captivity narratives are especially collaborative, she argues, because their female and minority authors are frequently reliant upon editors, translators, publishers, patrons, and others in order to reach publication. Logan asserts, "[t]o foreground collaborative authorship in the study of Indian captivity narratives is to emphasize the role of those historical and cultural processes through which these texts were produced" (466). This approach resonates strongly with my own methodology and choice of texts.

Other critics, however, argue quite strongly against such a positive analysis of collaboration. David Sewell claims "[t]here can be no written accounts by transculturated captives ... that make visible that imperceptible moment when the Other becomes I. [...] [W]hen ... transcribed and published by a literate editor, they are transformed into stories of *our* culture" (53). Sewell focuses on the artifact of the text itself, suggesting that only an oral retelling would suffice in the case of Mary Jemison, whom he names as his example. Stephanie Sellers employs a similar argument when she critiques Krupat's choices, such as categorizing *Life of Black Hawk* as Native American autobiography: "A better description of the Native American autobiographies of the 1800s would be Eurocentric, romanticized caricatures of Native Americans infused with some accurate retellings of Native history and personal Native narrative all presented in a western cultural context" (17). While she allows that Krupat's work is a worthwhile starting point, Sellers prefers that researchers work to create "Indigenous Communal Narratives." One example she cites is Paula Gunn Allen's 2004 biography of Pocahontas.[10]

Entering this lively and ongoing debate about Native American autobiography, I examine texts now seen as autobiography by some, but not originally labeled as such, as examples of the evolving genre of the captivity narrative. I argue that texts such as *Life of Black Hawk* and *A Narrative of the Life of Mrs. Mary Jemison* co-opted and manipulated generic conventions of the captivity narrative. We would do well to read these primary documents with rich contextual histories as captivity narratives. My evaluation of these texts as cross-cultural collaborations is based in part upon an acceptance of arguments by James Clifton (and Krupat) that Native American identity was based on language rather than ancestry. Clifton writes, "the most common identity question was not, 'What nation do you belong to?' or 'Of what race are you?' Instead, when confronting unknown people, they typically asked, 'What language do you speak?'" (11). Because captives like John Tanner were much more comfortable speaking Native languages than English, I consider such a text to be cross-cultural. While the idea of Indigenous Communal Narratives, for example, is yielding impressive results, I believe that reading these nineteenth-century primary sources as captivity narratives is both faithful and contextually accurate.

Collaborative texts are often discounted by critics who privilege the work of a single author as more literary or more authentic. Furthermore, captivity narratives have traditionally been seen as less literary than other nineteenth-century genres. This attitude has minimized the critical value of captivity narratives, which often have murky textual histories. Many texts have been republished with the editorial history omitted or otherwise obscured. This process continues today, even when editors identify the texts as American Indian autobiography. Some of the earliest critical work on these collaborative texts follows what Logan calls a "treasure hunt" approach: a search for clues to the author's authentic voice hidden within the editor's words (467). I argue that the collaborations and editorial choices made in the nineteenth century cannot be hidden or undone, nor should they be. The texts are remarkable without repackaging for a contemporary audience, and considerably more complex when read with an understanding of the motivations and methodologies of the editors. For example, many captivity narratives were published with lengthy appendices, which often included vocabulary lists in multiple Native American languages as well as natural history observations presented alongside accounts of Native American religious and cultural practices. These sorts of scholarly materials showcase early ethnographic work and add to the complexity and ambition of a text; it is a mistake to reformat and abridge in an attempt to focus solely on the biographical details of a captive.

Captivity narratives were published (and reissued) with pointed and timely goals. Mary Rowlandson's captivity narrative has been republished many times, often in conjunction with political movements. The 1770 edition argued for American independence from Britain in the form of its well-known frontispiece of Rowlandson firing a gun—an event that is historically inaccurate. In 1824, editor Joseph Willard reprinted Rowlandson's narrative with comments supportive of Indian relocation in the magazine *Collections Historical and Miscellaneous*. His commentary that accompanied excerpts from *The Sovereignty and Goodness of God*, as Rebecca Blevins Faery notes, argued for the text's renewed relevance because of several reasons: "to vilify Indians, to represent them as being still a terrifying threat to the peace and security of white Americans, to assert Euro-American 'rights' to the American continent, and to figure the American national subject as white" (57). On the other hand, Edwin James's 1830 introduction to *A Narrative of the Captivity and Adventures of John Tanner* explicitly argued against the proposed Indian Relocation Act, which was passed only months after the publication of the narrative.

Yet later editions of Tanner's narrative, such as James Macauley's 1883 *Grey Hawk*, reduce the accuracy of the text by omitting the introduction and the second part of the book. Indeed, one important reason to read various captivity narratives as the product of diverse editorial and authorial relationships is to examine how these cross-cultural relationships manifest in the text. Only occasionally do editor and captive have matching intentions for a collaborative text's argument and appeal, but the texts actively reinforce or challenge ideas of ethnic or national identity. By removing contextual and collaborative markers, authorial motivations are removed. The individual captivity narratives I focus later chapters upon were written to serve not as personal memoirs, but rather as passionate yet calculated attempts to influence land use decisions by arguing against Indian removal.

Captivity narrative scholarship

As I have mentioned, the Indian captivity narrative has only in passing been considered as a form of environmental writing, despite a major increase in literary scholarship on both captivity and environment. Study of early American nature writing has largely been confined to exploration and natural history texts such as William Bartram's *Travels* and the journals of Lewis and Clark. By the mid-nineteenth century, the literary nonfiction of such writers as Henry David Thoreau, Ralph Waldo Emerson, and Susan Fenimore Cooper had eclipsed other forms of environmental writing. Scholarship on these texts has increased in the past twenty years with the emergence of ecocriticism. Scholarship on Indian captivity narratives has similarly increased during the same period, but with a strong focus on race and gender. While this book surveys some of the possibilities that captivity narratives have for literary scholars interested in questions of nature and society in early American literature, a brief overview of captivity narrative scholarship shows that some groundwork has been laid.

The definitive study of captivity narratives as a genre is *The Indian Captivity Narrative 1550–1900*, by Kathryn Zabelle Derounian-Stodola and James Levernier, both of whom have written extensively in the field. Two of the chapter titles in this book serve as valuable shorthand for the two primary thrusts of this type of research: "Images of Women" and "Images of Indians." Gender and race are clearly identified as the twin focus of scholars of captivity narratives; my study expands this analysis by taking these ideas into consideration within the larger sphere of environmental justice concerns. Rebecca Blevins Faery's *Cartographies of Desire: Captivity, Race, and Sex in the Shaping of an American Nation* promises much in the way of mapping the frontier: her methodology includes "reading texts and landscapes by way of conventional textual analysis and personal narrative" (18). However, the reading of texts with concern for race and gender issues composes the book's analytical element, while landscapes are discussed only within narrative fragments that begin each chapter by describing the author's own research trips. "[T]he convergences of race and gender in the construction of America" thus remain privileged over the role of place (18).

While scholars of the captivity narrative have not fully engaged environmental questions, some have approached the questions I am most interested in. By the nineteenth century, the captivity narrative had broken away from overly prescriptive genre constraints, and environmental lessons and observations can more readily be found within it. Christopher Castiglia finds fictionalized captivity narratives such as Catherine Maria Sedgwick's 1827 novel *Hope Leslie* intriguing, noting that

> traditionally, the wilderness has been the domain of the American Adam. Literary criticism has neglected women writers in general but has been particularly reluctant to consider women in the wilderness. … When one considers the captivity romances, however, a different

> 'wilderness' emerges, and consequently a different configuration of gender outside the home (134).

For Castiglia, what makes writers like Sedgwick worthy of study is not the wilderness settings in their work, but rather the individual agency that her characters display by upsetting determinations of domestic womanhood that insist a woman's place is inside the home. Castiglia also briefly discusses nineteenth-century captivity narratives, including those by Rachel Plummer and Abbey Gardner-Smith, that contain detailed environmental descriptions and narrators who derive great strength and pleasure from the natural world.

Novels such as *Hope Leslie* and Lydia Maria Child's 1824 *Hobomok* are captivity romances, written even as the policy of Indian removal is being formed. Yet both of these novels are historical romances set in seventeenth-century Puritan communities, which seems to suggest that discussing contemporary conflicts between whites and Native Americans may result in similarly tidy endings. Hobomok yields his white wife to her former love interest, a white man who had been presumed dead. Hope Leslie, not Magawisca, marries Everell Fletcher. Sedgwick even opens *Hope Leslie* with the verses "For the white man came with power—like brethren they met— / But the Indian fires went out, and the Indian sun has set! // [...] Where has the rainbow vanished?—there does the Indian dwell" (1). The removal of Indians provides the framework of the novel, which is written by a woman considered to be among the most sympathetic writers of the period. The landscape is fully historicized—despite a rather radical and progressive revisioning of history—and does not allow for an alternative vision of what the contemporary or future environment might look like.

Michelle Burnham focuses on the idea of cultural exchange, using postcolonial insights, especially those of Homi Bhabha, to argue that colliding cultural differences can create a "supplement" in the liminal space between two cultures engaged in an exchange (*Captivity* 20). The captive, Burnham continues, can "contest and destabilize the presumed autonomy and homogeneity of monocultural systems" (21). Yet for Burnham, the space that is contested is above all a cultural and economic space. While she does look briefly at Andrew Jackson and Indian relocation, it is primarily to suggest that the debt owed to the dead, dying, and/or dispossessed tribes was "represented [...] as a spectacle of loss, which the imperial nation could only watch and mourn. The discourse of manifest destiny allowed the imperialist audience to paradoxically forgive their own debt" (*Captivity* 116). The land itself is simply the setting of the drama, important only incidentally in Burnham's monograph.

Even when ecocritics do examine captivity, their arguments have tended toward different theoretical approaches. Gordon Sayre's excellent work on captivity seems to be carefully compartmentalized apart from his ecocritical scholarship. For example, he has published on *A Narrative of the*

Captivity and Adventures of John Tanner, arguing persuasively that the text is presented as autobiography despite the difficulties of translation from one racially distinct culture to another. I argue the additional environmental point that the ability of these texts to bridge supposedly unbridgeable ethnic and social boundaries also speaks to the possibility of bridging false boundaries between civilization and the wilderness.

Once used both to preserve and to destroy the environment, captivity narratives remain a rich and underappreciated resource for critics interested in studying early American attitudes toward the natural world. An ecocritical perspective allows us to study this genre and its authors for their environmental influence during the dramatic expansion of the US in the nineteenth century. The popularity of the genre allowed many women and minorities to reach a wide audience. The liminal contact zone many critics identify as existing between white American and Native American cultures must also exist between differing landscapes that both sustain and are inscribed by their human inhabitants. Captivity narratives have much to offer an ecocritical analysis, as land use decisions made by different cultures profoundly impact nonhuman communities. For example, the burning of the Seneca villages, fields, and orchards in 1779 by Sullivan's Expedition radically altered the environment and inhabitants of the Genesee River valley, as depicted in Jemison's *A Narrative*. My later chapters therefore explicitly consider the preservationist legacy of specific captivity narratives.

Captivity narratives: Exploring an alternative environmental literary tradition

My second chapter examines *A Narrative of the Life of Mrs. Mary Jemison*, with particular focus on the collaboration between Jemison and the author of the book, James Seaver, as well as its late-nineteenth-century editor, William Pryor Letchworth. Jemison, who was captured by Seneca in 1755, married Seneca men and had mixed-race children. She refused in later years to return to white society; her story was recorded by James Seaver in 1824. The narrative was probably a regional text when first published, as Wayne Franklin argues convincingly.[11] Yet, the narrative's publication history shows lasting interest in Jemison's story. *A Narrative* was published twenty-seven times in twenty-three editions over the next 105 years (Namias, "Editor's Introduction" 4). Five of those editions appeared between 1877 and 1913, spearheaded by a wealthy Quaker businessman, William Pryor Letchworth (Namias, "Editor's Introduction" 41). Not only did Letchworth publish these reprints, adding to the contents of earlier editions, he also bought land along the Genesee River near where Jemison had lived, had her remains moved to the land, and commissioned a bronze sculpture of Jemison. Throughout these years, Letchworth worked with a landscape architect to restore the logged forests and dammed rivers (Beale 73–74, 95–96). Letchworth eventually donated his thousand-acre estate to the state of New York to become the centerpiece of a state park that today serves more than one million visitors annually, and is also the site of private powwows.

The story of Jemison is closely entwined with white projections of what "Indianness" meant. Jemison was reluctant to have her story recorded despite widespread curiosity about her from white settlers in New York; I argue that her eventual decision to talk to James Seaver was motivated by a desire to secure her land rights. Even more so than the other texts I examine, it is the product of its white editor/author James Seaver, and was later expanded by other editors. Nevertheless, Jemison's legacy is more impressive than that of other captivity narratives in terms of memorialization through land conservation, and the text has also undergone more reprints than virtually any other captivity narrative.

My third chapter looks closely at Edwin James as an important author who has been undervalued because of the collaborative nature of his works and his relationship with John Tanner. Tanner was taken captive as a child and adopted into an Ojibwe/Ottawa family; like Jemison, he has been labeled as a white Indian. Briefly and unsuccessfully reunited with his biological family thirty years later, Tanner found employment as an interpreter in the Michigan Territory where he met James. James's first major work, *Account of an Expedition from Pittsburgh to the Rocky Mountains* (1823), is remarkable for its preservationist view toward the Great Plains. *The Captivity and Adventures of John Tanner*, where James's most explicit comments on Indian relocation appear, makes a powerful case for Native American cultures and land rights. Trained as a botanist, James also draws out facts and stories about various plants and animals that are fascinating for environmental criticism today. Yet James and Tanner suffered from their competition with others in the Michigan Territory such as Henry Rowe Schoolcraft, and were on the losing side of political battles such as Indian relocation. Collaboration is of great importance in this chapter, as the nearly equal relationship between captive and editor is unique in the series of texts that I discuss. Indeed, the two men appeared as co-authors of a later text, a translation of the New Testament into Ojibwe. A lake along the border of Minnesota and Ontario, in the Quetico Provincial Park (a wilderness canoe area) has been named after John Tanner. Appropriately, Ojibwe peoples continue to use this lake for traditional purposes as well as when acting as fishing guides for white tourists.

The subject of Chapter 4, the story of the Sauk warrior Black Hawk, was published shortly after Black Hawk's defeat and subsequent incarceration by the US Army in 1832. Black Hawk's band crossed the Mississippi River from west to east, a symbolic event that triggered a military retaliation disproportionate to any offenses he may have committed. Most scholars consider *Life of Black Hawk* as autobiography and have focused on the idea of Black Hawk as a tragic hero. A famous portrait of Black Hawk by George Catlin helped create this legacy, but Catlin also painted a lesser-known group portrait that depicted the captive Sauks, including

Black Hawk, in chains. This later painting is more accurate, I believe, and suggests that *Life of Black Hawk* may be called a counter-captivity narrative rather than an autobiography. *Life of Black Hawk* invokes many common tropes of the captivity narrative, a genre with which the editors and publishers of the text were certainly familiar. While many captivity narratives were written in an attempt to justify the usurpation of Native American lands through the reinforcement of cultural assumptions about white superiority, *Life of Black Hawk* offers a parallel argument. Black Hawk argues for the superiority of the Sauk Nation, justifying his armed defense of their homelands and reflecting upon what has been lost. The project's editor, John Patterson, and also its translator, Antoine LeClaire, worked to retain Black Hawk's world view, as angry critiques of relocation practices and white settlers abound in the text. The textual history, then, reveals rhetorical resistance and survival despite an imbalance of power—what Gerald Vizenor calls "survivance." Black Hawk's home was at the mouth of the Rock River, which enters the Mississippi from modern-day Illinois. Today, this area has been preserved as Black Hawk State Historic Site and is home to outdoor education programs and a Sauk and Meskwaki (Sac and Fox) museum.

The final chapter in this book looks at the same captivity narratives from a communal perspective, rather than reading them as individual works for specific historical moments. The legacy of the texts is expanded and complicated by showing how present-day communities continue to use them in a range of places and ways. Finally, I also show an example of how communities with specific environmental and cultural goals and a history of captivity and exile are able to enact change through restorative practices, even when the specific landscape they associate with has been displaced and altered.

The three chapters that follow this introduction are focused on texts that were the products of cross-cultural collaboration. Because of their collaborative authorship, these captivity narratives are able to resist and even escape cultural and environmental stereotypes that too often marked texts of the early nineteenth century. Such collaborations offered a way for women and Native American writers along the contact zone of North America to make their voices more widely heard. Indeed, all of the texts Chapters 2–4 focus on have been recently been republished as autobiographies, which suggests the enduring strength of these voices. I believe, however, that a contextually based reading of these texts as collaborative captivity narratives offers a more accurate way to understand them. Through discourse rooted in particular landscapes and places, these texts helped to spur successful land preservationist efforts. Another motivating factor in the creation of the texts was the debate over Indian removal, and later, over the enforcement of such policies as the 1830 Indian Relocation Act. By decoupling race from culture, and more specifically, from a dichotomy of wilderness versus civilization, these subversive captivity narratives were able to challenge the supposed inevitability of the Vanishing Indian and westward expansion.

Notes

1. The Newberry Library has compiled a list of more than 2000 captivity narratives; this list does not include fictionalized accounts and is not considered exhaustive (Derounian-Stodola, Levernier 2, 8).
2. For example, many texts were published in the aftermath of the 1864 so-called Sioux Uprising and are detailed in Kathryn Zabelle Derounian-Stodola's *The War in Words: Reading the Dakota Conflict Through the Captivity Literature* (Lincoln: U of Nebraska P, 2009).
3. The full title of Ebersole's chapter is "Going Native, Going Primitive: White Indians, Sexuality, Power, and the Problem of Identity."
4. The story is: J. K. P. "The White Indian." *The Atlantic Souvenir; A Christmas and New Year's Offering* 1827. 56–95.
5. Similar in conception to Buell's appendix of "Environmental Nonfiction at the time of Thoreau's Emergence," Michael P. Branch's anthology *Reading the Roots: American Nature Writing before Walden* bears examination. Álvar Núñez Cabeza de Vaca's narrative (1542) and James Smith's *An Account of the Remarkable Occurrences in the Life and Travels of Col. James Smith* (1799) are the two excerpted texts included that fit a fairly narrow definition of Indian captivity. Regarding Smith, Branch writes that "in addition to being replete with detailed observations made in the deep wilderness, it [Smith's book] shows how his captivity is also a form of apprenticeship [...] with what we might call an environmental education" (191).
6. The minor discrepancies in capitalization, italicization, and punctuation are due to my usage of the second edition, rather than the fourth, which Mazel uses.
7. For more on Cynthia Ann Parker, see: Hacker, Margaret Schmidt. *Cynthia Ann Parker: The Life and the Legend.* El Paso: Texas Western Press, 1990.
8. This area, which she describes as being full of thorn-trees, none more than fifteen feet tall, is perhaps in Sonora, Mexico, but it is difficult to verify based on her description.
9. Krupat offers a helpful, though not exhaustive, bibliography of texts from 1820 to 1870 that deal with Indian removal in a direct manner (*Ethnocriticism* 57–71). The list includes drama, poetry, and fiction by such writers as Lydia Maria Child and James Fenimore Cooper, as well as many nonfiction texts, including a significant number of captivity narratives.
10. Allen, Paula Gunn. *Pocahontas: Medicine Woman, Spy, Entrepreneur, Diplomat.* New York: HarperCollins, 2003.
11. Confusingly, Frank Luther Mott claimed in 1947 that the text sold 100,000 copies by 1834. Wayne Franklin explains that this would have been impossible based on several factors, including the fact that small western New York publishers printed the book and would not have had the ability to print more than a few hundred, no national publications reviewed the book, and Seaver's obituary in 1827 didn't mention the book (16–19). Still, Jemison has been claimed as a major bestseller on the basis of Mott's statement by critics such as Namias and Derounian-Stodola ("The Captive as Celebrity" 78–79).

2 The great slide

Mary Jemison's ruptured narrative

A Narrative of the Life of Mrs. Mary Jemison upended the genre of the captivity narrative through its unusual story of a woman who refused to return to so-called civilization. *A Narrative* highlights Jemison's lifelong struggle to secure a domestic space in the wilderness. The composition of the text required collaboration with James E. Seaver, who recorded and published it under his name in 1824. Despite her pointed refusal at several points in the narrative to be "redeemed," Jemison's story was packaged as a captivity narrative by her collaborator and appealed to an expanding and enduring readership.

Mary Jemison was born at sea in 1743 and lived until 1833.[1] Her life was marked by many important events associated with the shift from America's colonial era to the early republic, a point Seaver labored to keep visible through his emphasis upon battles and dates. Her parents were Scotch-Irish immigrants who settled near what is now Gettysburg, Pennsylvania. When their settlement was attacked in 1758 during the French and Indian War, Jemison and most of her family were taken captive by Shawnees. Many of the captives, including her parents, were killed shortly thereafter because the group was closely pursued by a white militia. Mary Jemison was spared, however, and was soon sold to a Seneca family. She immediately underwent an initiation ceremony by which she was adopted. Jemison married a Delaware man named Sheninjee in 1760. After their first son was born, she famously made an arduous seven hundred mile journey from Ohio to New York, where she joined a major Seneca community along the Genesee River. Sheninjee died in 1762, to her great distress. Despite her repeated opportunities to return to white settlements, Jemison refused, citing her belief that whites would be prejudiced against her mixed-race children. About a year after Sheninjee's death, she married an older Seneca man named Hiokatoo. They had six children together over a marriage that lasted nearly fifty years, until his death in 1811. Their home was at Gardeau Flats, which Jemison owned for several decades—land that is now part of Letchworth State Park. Over a period of three days in 1823, Jemison agreed to meet with James E. Seaver, who had approached her with the intention of writing down her story (Namias, "Editor's Introduction" 3). After the resulting narrative was published, Jemison lived until 1833, eventually selling her land and spending the final years of her life at the nearby Buffalo Creek Reservation.

The success of Jemison's narrative created space for other collaborative captivity narratives, such as those of John Tanner and Black Hawk, to challenge Indian relocation more explicitly on a national scale in the early 1830s. Tanner's story is subversive thanks in no small part to the efforts of his white collaborator, Edwin James, who—like Jemison's collaborator, Seaver—wrote down the life story of an illiterate "White Indian." While Black Hawk's collaborators may have been less sympathetic than James, they felt secure in letting Black Hawk's defiant words stand with little comment, as Black Hawk's viewpoint was already well-known from the events of the Black Hawk War.

Seaver, however, was uncertain about how to proceed with Jemison's radical story. For this reason, *A Narrative* features an uneasy compositional collaboration. Telling her life story from a position of some strength as a major landowner, Jemison insisted on having her lawyer present during her interviews with Seaver. Although this strategy must have helped Jemison shape the first edition, improving her status in the region, the narrative went through more than twenty editions in the first century of its existence. Over time, editorial emendations and ancillary additions altered the shape of the text, making for a fascinating study of influence and reception. Charles Delamater Vail's "Tabulation of Editions," itself an artifact of longstanding editorial interest in the text, graphically illustrates the growth of the text from a compact, inexpensive book with no illustrations in 1824 to an ornate 483-page volume with forty-one illustrations in 1925 (Appendix 1).[2] Because of this ambivalent relationship between the historical figure of Mary Jemison and her many editors, various contradictions appear in the narrative—contradictions that highlight questions of genre, gender, and race. The compelling nature of these contradictions, and the importance and influence of the text in cultural and literary spheres, have led to a wide range of critical studies of the text. In addition to surveying the text's cultural and racial disruptions, which have been the focus of some critical attention, my chapter examines how Jemison's cautious collaboration with Seaver has important implications for the landscape. I argue that Jemison's major goal in agreeing to collaborate with Seaver on *A Narrative of the Life of Mrs. Mary Jemison* was to situate her domestic household in the wilderness through a variety of cross-cultural appeals, thereby establishing her land rights. While the collaboration was one Jemison approached with caution, she entered it willingly and the resulting text was valuable in publicizing her story and ensuring her legacy.

Later editions of *A Narrative* were published by the American Scenic and Historic Preservation Society, whose interest in Jemison as a preservationist of both cultural and natural resources led directly to important land conservation efforts. This work culminated in the 1906 establishment of a large state park that thwarted proposals that would have dammed the Genesee River and turned the "Grand Canyon of the East"—where Jemison lived for decades—into a reservoir. Hundreds of feet high, the canyon walls at

Gardeau give way to a major series of waterfalls just downstream and a gorge that, at some places, is more than five hundred feet deep. The Great Slide, mentioned in this chapter's title, is both a metaphor for Jemison's ruptured life and a reference to an 1817 natural disturbance caused by a dramatic landslide of twenty-two acres that altered the course of the Genesee River at the center of the land she called home.

"Taken by the Indians"

Michelle Burnham argues that while *A Narrative* is seen as a captivity narrative by those who consider Jemison to be white, it is seen as an American Indian autobiography by those who see her as a Seneca. For her part, Burnham claims that Jemison moves between these—and many other—fixed identities ("However Extravagant" 326). I argue that the captivity narrative is the most appropriate generic category for *A Narrative*. Rather than matching all the conventions of the captivity narrative genre, however, *A Narrative* is better described as a subversive captivity narrative. *A Narrative* exploits the expectations of the genre in order to make Jemison's unusual life story available to as wide an audience as possible, while altering the generic storyline in radical ways.

Seaver's 1824 first edition, published locally in Canandaigua, New York, promises much in the way of sensational stories of captivity. The book's original title immediately notes that Jemison was "taken by the Indians" and promises that the narrative will contain "An Account of the Murder of her Father and his Family; her sufferings; her marriage to two Indians" and "barbarities of the Indians" (5).[3] In his moralistic "Author's Preface" that opens the text, Seaver argues that "biographical writings" offer readers models by which "to select a plan of life that will at least afford self-satisfaction, and guide us through the world in paths of morality" (49). Seaver's framing of Mary Jemison's captivity narrative is much like Increase Mather's introduction to Mary Rowlandson's narrative: an explanation of what meaning readers should derive from the narrative they are about to read. Yet while some readers find Rowlandson subversive, Jemison is considerably more provocative in her choice to remain with the Seneca. Nonetheless, framing devices such as Seaver's preface and the conventions that Rowlandson helped create for the captivity narrative made the publication of Jemison's narrative more permissible.

Seaver followed the preface with his "Author's Introduction," writing that he "presumed"

> but few native Americans [white Americans] that have arrived to middle age, who cannot distinctly recollect of sitting in the chimney corner when children, all contracted in fear, and there listening to their parents or visitors, while they related stories of Indian conquests, and murders, that would make their flaxen hair nearly stand erect, and almost destroy the power of motion. (53)

This sensationalized yet conventional idea about the contents and effects of a captivity narrative is emphasized by Seaver, especially in the book's early chapters. After establishing the European origin of Jemison's biological family—and mentioning her birth aboard a transatlantic sailing ship, a kind of liminal space between the Old and New World—Seaver dwells at great length upon the traumatic details of Jemison's captivity:

> It is impossible for any one to form a correct idea of what my feelings were at the sight of those savages, whom I supposed had murdered my parents and brothers, sister, and friends, and left them in the swamp to be devoured by wild beasts! But what could I do? A poor little defenseless girl (70).

On the following page, Jemison relates seeing the scalps of her parents "yet wet and bloody" (71).

Thus, the text begins with the exploits of the "cruel monsters" that Seaver presumes his readers desire (71); yet Seaver is unable to sustain this stylized and vitriolic tone because at several points in the narrative his subject refuses to rejoin white communities. Some critics have categorized Seaver's strategy as a failure. Hilary E. Wyss states that "Seaver's narrative seems to constantly slip from his control; try as he might to shape it into a form with which he is comfortable, the details of Jemison's experience elude clear narrative boundaries" (67). Readers expecting a conventional captivity narrative in which the captive is providentially delivered to civilization are disappointed. Due to this thwarted genre convention, Wyss argues, the "form of the captivity narrative imposed by Seaver, then, is inadequate to contain Jemison's story" (69); a later sentence notes that "Jemison's is a 'failed' captivity narrative that shifts the text to something more akin to a conversion narrative" (69). Susan Walsh likewise argues that the conventional story failed after only a few chapters: "once Jemison is adopted as a Seneca, her story spills over the sides of its literary container so that what had begun as a melodrama of beset womanhood segues into a tribute to departed sisters, brothers, husbands, and children" (54). While *A Narrative of the Life of Mrs. Mary Jemison* is presented by Seaver in the early going as a conventional captivity narrative, the facts of Jemison's long life ultimately made the formulaic structure Seaver attempted to impose on the narrative unconvincing.

Yet categorizing *A Narrative* as a "failed" captivity narrative ignores the more compelling and valuable appeal of *A Narrative*. The captivity narrative genre was particularly useful for Jemison as she was able to gain a wider audience than would have been possible had not the text carried this label. In subverting conventions, Jemison does not abandon the genre, but rather is able to complicate and disrupt the language of captivity in order to shift reader expectations. Seaver's role in the collaboration should not, in my view, be labeled as exploitation; indeed, Jemison's authentic and compelling story repeatedly overwhelms the expectations and assumptions Seaver attempts to impose upon it.

Specifically, Jemison refused to be "redeemed" through a return to Christianity or white culture. Laura Mielke summarizes the critical work done on this point, writing that *A Narrative* "does not include the crucial ending of the captivity narrative, the return and reincorporation of the captive, and therefore it remains on the margins of the genre" (79). One of the most memorable episodes in *A Narrative* occurs during the short time Jemison is a widow in 1762, when a bounty is offered for white captives. First, she physically outruns a man named John Van Sice who attempts to capture her. Van Sice's efforts inspired a Seneca leader to earn the bounty, however, and he attempts to take her to Niagara with other whites; her adoptive brother vows to kill Jemison rather than letting the chief take her away. Jemison and her sister secretly devise a warning, should the threats be attempted against her. Jemison relates: "If I was to be killed, she [my sister] would bake a small [corn] cake and lay it at the door [...] if the cake was there, I was to take my child and go as fast as I possibly could to a large spring" (94). Jemison thus becomes a fugitive until she is assured she will not have to return to the whites, and the inclusion of corn, which is held sacred by the Seneca, is an important symbol of her acculturation. As Walsh points out, the cake also holds important currency in the domestic realm, making it a highly charged symbol (55). For Jemison as an adult and mother, the threat of captivity remained, but was represented by the whites.

In addition to refusing to return to white settlements, Jemison rejects Christianity during her adult life, unless a supposed conversion in the final months of her life in 1833 is accurate. On the issue of her faith, Seaver notes in the introduction that "[h]er ideas of religion, correspond in every respect with those of the great mass of Seneca" and "[t]he doctrines taught in the Christian religion, she is a stranger to" (58). Yet he writes in Jemison's life story itself how her mother instructed her to faithfully recite her prayers; as her mother is led away to her presumed death, she says, "Don't cry Mary—don't cry my child. God will bless you!" (69). This romantic and sympathetic vision of the bond between mother, child, and God is maintained and developed by Seaver and later editors, who found this trope extraordinarily compelling. These editors worked to bend the arc of Jemison's faith as closely as possible to match the formula of a redeemed captive. While this work may have created a text that grew more conventional over the years, it also helped ensure a lasting appeal to a wide readership that continued to be exposed to the underlying, subversive claims of *A Narrative*.

By 1842, James Seaver had been dead for nearly fifteen years. At this point, his brother William Seaver collaborated with Ebenezer Mix to bring a new edition to press. This edition made many changes to the text, including the reorganization of the text with silent omissions and additional chapters on geography and history. Finally, the William Seaver and Mix editions introduced the claim that Jemison became a Christian in 1833. A yet later editor, William P. Letchworth, continued this editorial trajectory by expanding further upon the supposed conversion of Jemison. Letchworth's motives, often pursued at personal expense, were more progressive than Mix's, but it is important to detail how he added some of the most melodramatic passages to the text. In the 1877 edition, Letchworth published a letter about Jemison's religious conversion by Laura Wright, who was the wife of Reverend Asher Wright and had known Jemison at Buffalo Creek Reservation.[4] William Clement Bryant (distinct from William Cullen Bryant) claimed "few hearts [are] so hardened as to be unmoved by the matchless pathos of Mrs. Wright's narrative" (200). The letter is five pages long and describes how Wright ministered to Jemison's frantic and even pathetic desire in 1833 to remember the prayer her mother told her upon their separation seventy-five years earlier. Most critics are dismissive of Wright's story, in no small part because Wright's letter was written more than 115 years after Jemison's mother supposedly directed Jemison to pray daily. In any event, Wright opines that

> I thought it a remarkable instance—the permanent influence of a mother's teaching. ... there is good reason to believe that she died in the cheering faith of the gospel, and not in the darkness of paganism, by which she had been for so many years surrounded (211–12).

Wright also connects the restoration of Jemison's religion to her color: "she had a very white skin ... Her face was somewhat bronzed by long exposure; but I noticed that at the back of her neck her hair was a bright color and curly, and her skin very white" (212). As Mielke summarizes, "Wright brings *Narrative* to its generic fruition, granting the captive's religious redemption in a moving dramatization of the victory of her 'civilized' identity despite years of residence among the Indians" (83). Notably, Wright's chapter is titled "Last Hours of the Captive," which insistently restores Jemison's status as a captive, despite her choice at age ninety to live on the Seneca reservation.

Contemporary unwillingness to grant the deathbed conversion story by Laura Wright much credence also has origins in the 1824 edition of *A Narrative*. In that edition, the text states plainly that "the attempts which have been made to civilize and Christianize them [the Indians] by the white people, has constantly made them worse and worse" (84). Seaver attempts to solve this inconsistency by allowing Jemison to be identified with both whites and Native Americans. While Jemison herself seems to identify primarily with the Seneca, the text here oddly segues to claim how failed attempts to "civilize and Christianize them [Indians] ... will ultimately produce their extermination" (84). Seaver's ambiguous use of pronouns, while consistently telling the story from Jemison's first-person point of view, allows him to slide Jemison from her Seneca identity to a white identity when he laboriously adds conventional language to her otherwise unconventional story.

Laura Wright's insistence that Mary Jemison had white skin is also telling. Various editors emphasize Jemison's appearance, suggesting that despite her own clear choice to remain with the Seneca Iroquois, her white editors did not consider her an Indian. Michelle Burnham argues that "[e]ven while detailing her Indianness, however, Seaver insistently identifies Jemison as white and distinguishes her both racially and culturally from the Indians" ("However Extravagant" 326). In his introduction, Seaver calls Jemison "The White Woman" (56), a name that she had already been well known by in the popular regional press (Franklin 18). This appellation persisted, finding its way onto the title page of many later editions. A bit awkwardly to a reader today, Seaver in the same page describes her clothing as being "after the Indian fashion" (56). Ezra Tawil writes that Jemison's narrative "insisted to her readers that she remained 'white'" despite marrying Seneca and having children and thus the story asserts her whiteness and race to be "something that could not be lost or taken away" (102). Rather than depicting cultural differences by national or religious characteristics, Jemison's narrative "represented race as an unbridgeable natural difference" (Tawil 102–03). Seaver's writing thus casts Jemison in the heroic role of a domestic white woman who accepts her trials willingly. Furthermore, Seaver can be seen as attempting to make Jemison an unusual model of Republican motherhood in that she keeps her mixed-race children away from white children, rather than remaining with the Seneca out of preference. This view, which Seaver may well have held, suggests that Native Americans cannot become civilized and whites cannot become "Indianized," even after a lifetime together and when bound by familial ties. Seaver's contortions are unconvincing today, and likely were unsatisfactory even in 1824. Nonetheless, he was working at a time when scientific racism was on the rise, leading to a popular belief in phrenology and polygenism; as Ian Finseth explains, this latter theory of multiple human origins "took the idea of race as spiritual essence to its extreme, denying the power of environment to create, sharpen, or minimize racial difference" (142).

Seaver's viewpoint suggests that Indian relocation is necessary; his argument, also drawing on the idea of the Vanishing Indian, proved very influential, even for later generations with more sympathetic and culturally sensitive outlooks. In 1877, the same year Wright's letter was published by Letchworth in the new edition of *A Narrative of the Life of Mary Jemison*, Letchworth's close friend David Gray published a related article in *Scribner's Monthly* entitled "The Last Indian Council on the Genesee." Gray's title, which echoes James Fenimore Cooper, is heavily reinforced throughout the article. He quotes Mary Jemison's grandson Thomas as saying

> It is painful to think that in the course of two generations there will not be an Iroquois of unmixed blood within the bounds of our state; that our race is doomed, and that our language and history will soon perish from the thoughts of men. But it is the will of the Great Spirit, and doubtless it is well. (346)

Rather than pointing out the oddity of such a quotation coming from the grandson of a person whom Gray calls "The White Woman," Gray instead suggests that Thomas Jemison's speech "is the voice of a moribund nation in the agonies of dissolution" (347). Of course, Mary Jemison was adopted as a member of the Seneca, which points to the fact that the Seneca were more concerned about their cultural identity than a racialized identity. The fact that this gathering occurred many decades after the Indian Relocation Act was passed—and that more than two dozen Iroquois were in attendance—suggests that Gray was hewing more closely to romantic convention than to reality. Seaver's contortions meant to establish Jemison's whiteness are likewise unconvincing.

Evolving views of the wilderness

Seaver's attempt to create a formulaic captivity narrative was stymied by the reality of Jemison's familial account, which allowed him to place her in the "dark and dismal swamp" for only a short while (68). Annette Kolodny notes that the various sentimental ploys used by Seaver to create a conventional heroine were "devices" meant to "prepare his audience for what he knew to be a most unusual text; but it suggests also that he himself may not have been fully prepared for the narrative he received" (73). Jemison is, according to Kolodny, "at home in the wilderness," but, in contrast with Daniel Boone, she "never achieved mythic status" (xiii).[5] While the opening chapters of *A Narrative* are notable mainly for Seaver's attempts to maintain a conventional storyline, Jemison's adoption ceremony is a major turning point.

The trajectory of Jemison's life takes a sharp turn once Seaver can no longer place her with her original Shawnee captors. As related in *A Narrative's* Chapter 3, Jemison is adopted by a Seneca family that has been in mourning for a brother killed in battle:

> His spirit [the brother's] has seen our distress, and sent us a helper whom with pleasure we greet. Dickewamis has come: then let us receive her with joy! ... gladly we welcome her here. In the place of our brother she stands in our tribe (77).

Jemison, known from this point on as Dehgawanus or Dickewamis, became thoroughly a part of her Seneca family. Her new name, translated as "Two Falling Voices" or "Two Wailing Voices," Burnham notes, "captures not only her own bilingualism but the complex bivocalism" of the narrative itself ("However Extravagant" 326–27).[6]

After her adoption, Jemison describes life for the Seneca in a village called Shenanjee, on the Ohio River. The landscape descriptions in *A Narrative* veer quickly from the swamp of her first night of captivity to a pastoral home: "the land produced good corn; the woods furnished plenty of game, and the waters abounded with fish. … [W]e planted, hoed, and harvested a large crop of corn, of an excellent quality" (79). Winters were spent downstream where better hunting opportunities existed: "forests on the Sciota [River] were well stocked with elk, deer, and other large animals; and the marshes contained large numbers of beaver, muskrat, &c." (79). These descriptions of daily life are more complete and spontaneous than those describing her life prior to captivity. For example, here is an earlier description:

> My father, with the assistance of the oldest sons, repaired his farm as usual, and was daily preparing the soil for the reception of the seed. His cattle and sheep were numerous, and according to the best idea of wealth that I can now form, he was wealthy (64).

Her language is more passive in this pre-captivity passage, as she seems to take the role of tour guide to what was around her, rather than describing her own role in the production of food.

In Jemison's new life, which within two years included a husband and child, she notes that after four years "with them was my home; my family was there, and there I had many friends" (83). Her wilderness experience was transformed into an experience of domestic family life and agricultural labor. Rather than finding her life to be full of drudgery, she argues that "[o]ur labor was not severe; and that of one year was exactly similar … to that of the others" (83–4). Jemison adapted quickly to what Kolodny calls the "cyclical idyll of Indian life" (77). In this section of the text, Jemison details many aspects of Seneca life, such as farming and cooking methods, and critics have thus found the narrative to be of ethnographic and environmental value—one reason why it is not "only" a captivity narrative. Indeed, the sympathetic anthropologist Lewis Henry Morgan edited the 1856, 1859, and 1860 editions of *A Narrative*, adding a lengthy appendix listing Iroquois place names across New York and into Canada. Karen Kilcup directly asks, "What do we ultimately learn about the environment from Jemison's narrative? […] Perhaps the story's most striking rhetorical feature … the corn-centered language, providing cyclical coherence and obliquely but powerfully resisting white society" (68). The emphasis on community life and the seasonal cycles has much resonance for ecocritics.

One major break in the ritual of the seasons occurs, and it is an episode that has been of enduring interest to editors and readers of *A Narrative* due to its heroic appeal. In 1762, Jemison and her son, who was younger than one year old, accompanied her three brothers in a move from the Ohio River to what they called Genishau (Genesee River area) in the fall, under the immediate influence of the major Seneca settlement most often called Little Beard's Town. Genishau, according to Jemison, "signifies a shining, clear or open place" (86); Morgan translates Genesee as the "Beautiful Valley" ("Indian" 266). Interestingly, Jemison notes that her brother insisted "upon our going home (as he called it)" (87). Jemison's sisters had already been living back in Genishau for two years at this point. She details some of the challenges of the journey "home," explaining the time of year based on the seasons and what food was available along the way. The trip was difficult, as frost killed much of the corn that year, and they suffered through many days of cold, drenching rain. While the episode occurs before Jemison ever sees the Genesee River valley, at the telling of her story she had been a resident of the area for decades. By emphasizing that this place was home—whether the parenthetical aside was hers or Seaver's—Jemison is laying advance claim to the area.

Jemison's endurance and strength as a mother are celebrated in this episode. If critics wish Jemison to be remembered as a hero, inspiration may come from these lines: "Those only who have travelled on foot the distance of five or six hundred miles, through an almost pathless wilderness, can form an idea of the fatigue and sufferings that I endured on that journey" (88). In addition to the distance, Jemison explains that her clothing and shelter was insufficient against the weather, and earlier she had described the difficulties in finding food along the way. Even more impressive, Jemison points out that she was a young mother at the time of the journey: "I had to carry my child, then about nine months old, every step of the journey on my back, or in my arms, and provide for his comfort and prevent his suffering, as far as my poverty of means would admit" (88). Mirroring her delivery from the Shawnee captors into her adoptive family, the trip to Genishau ends with a reunion with her Seneca sisters: "I am constrained to believe that I loved them as I should have loved my own sister had she lived, and I had been brought up with her" (89). This line appears at the end of Chapter 4 of *A Narrative*, and shows the connection Jemison makes between family and home. By conflating the land with family, she challenges nineteenth-century ideas of womanhood and the frontier, creating a domestic sphere in the midst of a supposed howling wilderness. This strategy is used regularly in *A Narrative*, and the connection between family and land is made time and time again in the concluding paragraph of later chapters.

While some time at Genishau was spent evading people who wished to return her to the whites, Jemison also mourned the death of her first husband, Sheninjee, who died before he was able to rejoin his growing family. She was soon remarried, this time to a significantly older Seneca warrior, Hiokatoo, with whom she had six children.[7] Hiokatoo is rarely mentioned by Jemison. Indeed, for the details of his life Seaver depends upon an unreliable man who claimed to be George Jemison, a cousin of Mary Jemison, and who eventually swindled Mary Jemison out of several hundred acres of land. Despite his uncritical acceptance of George Jemison as a trustworthy source, Seaver does add a qualifying footnote stating that "Mrs. Jemison is now confident that George Jemison is not her cousin, and thinks that he claimed the relationship, only to gain assistance" (144). Instead, the "twelve or fifteen years" (96) between wars, which ends with the Revolutionary War in 1776, is described as a time of peace and happiness. Jemison notes both spiritual practices as well as "various athletic games" to help show the range of cultural traditions (97). Men hunted, while "women attended to agriculture, their families, and a few domestic concerns of small consequence, and attended with little labor" (97). Once again, Jemison's depiction of peacetime life for the Seneca emphasizes the domestic and the cultural.

The consistent means by which Jemison establishes residency through agricultural practices can be read not only as a recounting of seasonal rituals, but also as a conscious effort to assert her land rights. Morgan, who was responsible for three editions of *A Narrative* and conducted research with the Seneca lawyer and engineer Ely S. Parker, opens Chapter 4 of *Houses and House-life of the American Aborigines* (1881) with an explanation of Iroquois theories of land ownership:

> Individual ownership, with the right to sell and convey in fee-simple to any other person, was entirely unknown among them. … No person in Indian life could obtain the absolute title to land, since it was vested by custom in the tribe as one body; and they had no conception of what is implied by a legal title in severalty with power to sell and convey the fee. But he could reduce unoccupied land to possession by cultivation, and so long as he thus used it he had a possessory right to its enjoyment which would be recognized and respected by his tribe. Gardens, planting-lots, apartments in a long-house, and, at a later day, orchards of fruit were thus held by persons and by families. Such possessory right was all that was needed for their full enjoyment and for the protection of their interest in them. (79)

With this understanding of land rights in mind, Jemison's detailed recounting of planting corn, squash, and beans may be understood not only as a description of daily life, but as an attempt to prove her right to remain at Genishau, as well as her right to have lived earlier at Shenanjee and Sciota. However, this mode of establishing land rights was challenged by white Americans during the Revolutionary War, in which the Seneca sided with the British.

In 1779, Sullivan's Expedition destroyed Jemison's village, along with dozens of other Iroquois communities. John Sullivan followed closely George Washington's directions to conduct a scorched earth policy against the Iroquois: "The immediate objects are the total destruction and devastation of their settlements, and the capture of as many prisoners of every age and sex as possible. It will be essential to ruin their crops now in the ground and prevent their planting more" (Washington, 31 May 1779). For the Seneca, this was a devastating period. Jemison's pronouns here identify fully with the suffering of the Seneca: "A part of our corn they burnt, and threw the remainder into the river. They burnt our houses, killed what few cattle and horses they could find, destroyed our fruit trees, and left nothing but the bare soil and timber" (104). Remarkably, Jemison was able to provide for her five children by husking corn for two fugitive slaves who lived at the Gardeau Flats, which she reports was within a day's walk of Genishau. Jemison suggests she found some humor in the fact that the blacks worried she would "get taken off or injured by the Indians" (105). Despite her ability to provide for her family, the following winter was a time of great trial, as Jemison notes that "snow fell about five feet deep, and remained so for a long time, and the weather was extremely cold; so much so indeed, that almost all the game upon which the Indians depended for subsistence, perished, and reduced them almost to a state of starvation" (105). Again, the pronouns shift. Seaver may be unable to decide upon a cultural identity for Jemison, whom the blacks apparently saw as a white refugee. Or, perhaps Jemison herself was exploiting her liminal status and is here able to instantly shift from one racial category to another depending on the situation, and the grammar of *A Narrative* reflects her canny decisions. For their part, the blacks later moved to "a place that they expected would suit them much better" (106). Yet Jemison remained at the Gardeau Flats from 1779 to 1831. While she managed to remain on the land despite the devastation wrought by Sullivan's Expedition, waves of challenges to her land rights threatened her over the following decades. Amazingly successful in most instances, Jemison's attempts to safeguard her home at Gardeau during the period of the early American republic came at a high price.

The great slide

In *A Narrative*, Jemison offers a precise description of her home:

> The land which I now own, is bounded as follows:—Beginning at the center of the Great Slide and running west one mile, thence north two miles, thence east about one mile to Genesee river, thence south on the west bank of Genesee river to the place of beginning (156).

Her land originates "at the center of the Great Slide," and the description of her two square miles of land is carefully rendered in the terms of legal boundaries. While the Great Slide is mentioned several times in the narrative, at this point in the text Seaver added the following descriptive footnote:

> The Great Slide of the bank of Genesee river is a curiosity worthy of the attention of the traveller. In the month of May, 1817, a portion of land thickly covered with timber, situated at the upper end of the Gardow [sic] flats, on the west side of the river, all of a sudden gave way, and with a tremendous crash, slid into the bed of the river, which it so completely filled, that the stream formed a new passage on the east side of it, where it now continues to run, without overflowing the slide. This slide, as it now lies, contains 22 acres, and has a considerable share of the timber that formerly covered it, still standing erect upon it, and growing. (156)

Seaver's footnote suggests his discomfort with Jemison's legalistic description of the area. Rather than altering the words in the narrative itself, he is compelled to add a footnote, a strategy he rarely used in the text (the handful of footnotes are most often in the service of correcting Jemison's dating of battles and other events). That Seaver felt unable to add his own interpretation directly into the text at this point—despite his willingness to claim authorship of the text as a whole—is likely due to the influence of Jemison's lawyer, Thomas Clute. Seaver's introduction explains that Jemison was "so jealous of her rights, or that she should say something injurious to herself or family, that if Mr. Clute had not been present, we should have been unable to obtain her history" (57). Not coincidentally, the Great Slide is also mentioned on the same page of the introduction: "Her dwelling, is about one hundred rods north of the Great Slide, a curiosity that will be described in its proper place" (57). While Seaver ostensibly is calling the Great Slide a curiosity and explaining that he will write about it later, the irregular grammar of the sentence argues instead that Jemison's domestic sphere in this rugged landscape is the true curiosity. Seaver's need for an authentic subject underlines that the "proper place" for Jemison to live is along the Genesee River, although he may accept this right grudgingly.

The Great Slide, as Walsh notes, is used elsewhere in the text to show Jemison's experience as a white woman who became an Indian woman. Within the exposed soil were human bones of unknown origin, and they point to "a cycle of displacement borne out by the Senecas' wartime dispersal and Jemison's provisional tenancy of her flats" (Walsh 64). The bones and the Great Slide greatly disturbed Jemison's oldest son by Hiokatoo, John, who was a well-known healer who treated Native Americans "by the administration of roots and herbs, which he gathered in the forests, and other places where they had been planted by the hand of nature" (147). For John, the Great Slide was an omen that foretold his impending death—which indeed came within days, as he was killed in a drunken quarrel. For Mary Jemison, this was only the latest tragedy, as her family was rapidly dissolving under the stresses of alcohol and white encroachment. Shockingly, John had already murdered two of his own brothers, Thomas and Jesse. Thomas had long criticized John's practice of polygamy and also considered him to be a "witch" (124); both Thomas and John were prone to violence while drunk. Jesse was a target because he "shunned the company of his brothers, and the Indians generally, and never attended their frolics" (142). Hiokatoo had also passed away, reportedly at age 103. While the death of John caused Mary Jemison sorrow "as a mother," she admitted that "I could not mourn for him as I had for my other sons, because I knew that his death was just, and what he had deserved for a long time" (149). Despite these events, Jemison remained at the Gardeau Flats, and emphasized her ability—and that of her daughters—to manage their large holdings.

In order to maintain her rights to the Gardeau Flats, Jemison pursued a variety of tactics. She followed traditional Seneca traditions, attended a treaty signing, drew up legal contracts with white settlers, and eventually even became a naturalized US citizen. Jemison's efforts to solidify her claim to the land were remarkably successful. Yet Jemison did not live at Gardeau Flats for the remainder of her long life. About eighty-two years old when *A Narrative* was published in 1824, Jemison lived nine more years. When she was about eighty-eight years old, she sold her land and lived at the Buffalo Creek Reservation with her family. The exact motivations for Jemison's move to the Buffalo Creek Reservation in 1831 remain unclear, but Namias lists many possible factors, suggesting that a combination of "various frauds, land sales, and her own old age, convinced her and her daughters to sell the remaining plots and move" ("Editor's Introduction" 33). Ebenezer Mix pointed out in the 1842 edition that all other Seneca land holdings on the Genesee were sold in 1825. He concludes that the Jemisons were "surrounded by the whites" and desired to "join their tribe, mix in the society, and partake in the joys and sorrows of their kindred" (194). While Jemison died in 1833, the Buffalo Creek Reservation survived only a few years longer. The 1838 Treaty of Buffalo Creek, which was widely denounced by the vast majority of the thousands of residents, was intended to force the Native Americans to remove to what is now Kansas. In this trajectory, Jemison's own experiences can be seen as a kind of case study for Native American land rights as a whole. The publication of *A Narrative* should also be seen as a way for her to plead her case, which refers to an important feature of the genre of the captivity narrative.

Jemison's original strategy to prove her land rights seems to have been a variation on the Seneca tradition described in Morgan's summary. Just as she did at Shenanjee, Sciota, and Genishau/Little Beard's Town, Jemison begins by describing her work on the land. First she labors in existing fields for the fugitive slaves, who were harvesting their crop. When they leave, she assumes ownership. In an attempt to head off interlopers, she enlists the authority of Seneca elders: "My flats were cleared before I saw them: and it was the opinion of the oldest Indians that were at Genishau, at the time that I first went there, that all the flats on the Genesee river were improved before any of the Indian tribes ever saw them" (106). She thus argues that nobody within human memory has preexisting rights to these lands. Yet she also invokes geologic time that echoes the Great Slide:

> I well remember that soon after I went to Little Beard's Town, the banks of Fall-Brook were washed off, which left a large number of human bones uncovered. The Indians then said that those were not the bones of Indians ... but that they were the bones of a race of men who a great many moons before, cleared that land and lived on the flats (106).

Kolodny writes that "the impact is immediate and undeniable: the uninhabited flats become a part of the human world, and the reader thereafter ceases to think of Jemison as living in an unremitting wilderness" (79). Causing the reader to think of Jemison as living in a domestic sphere, rather than the wilderness, is of course critical to Jemison's strategy of collaborating with Seaver in order to show that she has the right to remain at the Gardeau Flats. By invoking prehistory, Jemison argues that the wilderness has been actively inhabited—and "improved"—for thousands of years. This surprising argument directly contradicts myths of the wilderness suggesting that only whites used the land to its "full potential" through agrarian practices; such rhetoric was used to justify Indian removal in the debates raging in 1824 and culminating in the Indian Relocation Act of 1830.

Jemison's strategy succeeded for decades, as she was able to build log homes at Gardeau and raise her children to adulthood without incident after the Revolutionary War. White encroachment eventually resulted in the need for formal treaties, however. Jemison actually refused one more time to return to white settlements after the war—this some twenty years after her initial abduction—telling her adoptive brother Kaujisestaugeau that she intended to remain in so-called captivity for her entire life. Jemison relates that "he informed me, that as that was my choice, I should have a piece of land that I could call my own, where I could live unmolested, and have something at my decease to leave for the benefit of my children" (120). The poignancy here is due in no small part to Jemison's age at the time of the telling of her story. She was eighty-three and had many living descendants, a point with which she concludes the book: "I have been the mother of eight children; three of whom are still living, and I have at this time thirty-nine grand children, and fourteen great-grand children, all living in the neighborhood of Genesee River, and at Buffalo" (160). Issues of place in the conclusion, as elsewhere in the narrative, are closely bound with ideas of family.

Kaujisestaugeau, however, was only one of the Seneca leaders with the authority to assign land claims; others, such as Red Jacket, opposed Jemison's claim to the Gardeau Flats.[8] In order to solidify and formalize her claim, she attended the 1797 Treaty at Big Tree. This important council was attended by three thousand Seneca, and saw the establishment of many reservations while also legally opening up settlement for whites. Jemison relates that a Seneca named Farmer's Brother or Honayewus "presented my claim" (121). Jemison had talked to Farmer's Brother, and "I accordingly told him the place of beginning, and then went round a tract that I judged would be sufficient for my purpose, (knowing that it would include the Gardeau Flats,) by stating certain bounds with which I was acquainted" (120–21). Jemison's successful request in the Treaty is documented as follows:

> [O]ne other piece or parcel at Gardeau, beginning at the mouth of Steep Hill creek, thence due east until it strikes the old path, thence south until a due west line will intersect with certain steep rocks on the west side of Genesee river, then extending due west, due north and due east, until it strikes the first mentiones bound, enclosing as much land on the west side as on the east side of the river. ("Agreement" 1029)

In stark contrast to the treaties by which whites cheated Native Americans out of ancestral land, Jemison's claim at the Treaty of Big Tree was entirely successful. By using landmarks that she was much more familiar with than were the white negotiators, she was able to retain 17,927 acres—more than twenty-eight square miles, solely reserved for her family—much of it good bottomland well-suited to agriculture. This favorable outcome was only possible due to Jemison's willingness to collaborate with negotiators—including Farmer's Brother and others—and her knowledge of the landscape. By understanding the expectations of her audience, both in this instance and in the creation of *A Narrative*, Jemison shows an ability to manipulate genres and modes of communication for her own interests.

Once her claim was secured, Jemison admitted that while the land was excellent, it "needed more labor than my daughters and myself were able to perform ... The land had lain uncultivated so long that it was thickly covered with weeds ... I accordingly let it out [leased it], and have continued to do so, which makes my task less burthensome" (122). This description of Jemison's landholdings concludes the chapter, and thus seems to close the question of her right to live at Gardeau Flats. The following chapter is focused entirely on family, which follows the pattern of connecting land to the domestic sphere.

Once again, Jemison's domestic tranquility failed to last; in *A Narrative*, Chapter 10 describes the end of the interlude marking Jemison's happy occupancy of Gardeau. Jemison's problems in the period from 1797 to 1824 arise primarily from within the family. Earlier in her life, war and other outside factors forced her to move. During this later period of upheaval, Jemison's place of residence does not change, but rather is slowly degraded and reduced in size alongside the loss of family members—particularly the male members of her family. These losses included the natural death of her husband and the violent deaths of three sons mentioned earlier, as well as the eventual disavowal of her alleged cousin, George Jemison. George Jemison entered Mary Jemison's life in 1810, and she allowed him to live in a house she owned because of his poverty and her strong belief in hospitality and kinship. She notes that "I supported [George] Jemison and his family eight years, and probably should have continued to do so to this day, had it not been for the occurrence of the following circumstance" (144). At great length, Mary Jemison explains how George and another person convinced her to sign a deed describing land she would grant to them. She explicitly notes that she attempted to delay until Thomas Clute (her lawyer and neighbor) could return, but they prevailed. In the treaty of 1797, she controlled the terms of description, using local terms for land formations. Yet George's deed is drawn up using chains and links as units of measurement, terms with which she was unfamiliar. Assured that she was only granting the use of two fields that totaled forty acres, she agreed, unaware that the total was instead four hundred. Sometime shortly after the land deal, George beat her grandson for attempting to retrieve a cow, and Mary Jemison's limits were reached. She notes that "I got him [George] off my premises as soon as possible" (146).

After these problems from a supposed cousin, Jemison was wisely skeptical of her neighbor Micah Brooks, who approached her in 1816 and "observed that as it [her land] was then situated, it was of but little value, because it was not in my [Jemison's] power to dispose of it" (153). Brooks offered to help her become naturalized as a US citizen, after which her land could be assigned legal title—and presumably he could make an offer on a parcel. As Jemison had long been leasing her land, she had a clear idea of what the land might be worth to her financially. By this time, the other Seneca were already scattered on various reservations and small land grants, a loose network of which Gardeau was technically a part. With her husband deceased, and she already being quite elderly, Jemison was interested by Brooks's suggestion. Once she arranged for Thomas Clute to go over the details with Brooks, Clute confirmed that because "the sale of Indian lands, which had been reserved, belonged exclusively to the United States, an act of the Legislature of New-York could have no effect in securing to me a title to my reservation" (154). All parties then agreed that Jemison would need to become a US citizen to legally own her land; her naturalization was achieved in April 1817. This led rapidly to a series of sales that greatly reduced the size of Jemison's holdings. Jemison soon executed her deed and leased most of her land to Brooks and Jellis Clute in August of the same year (154–55). She also granted Thomas Clute "such a piece as he chose" (155), while reserving 4000 acres for her family's use. In 1823, she sold most of the land to Brooks, Jellis Clute, and H. B. Gibson in exchange for the payment of "three hundred dollars a year forever" to Jemison and her heirs (156).

Once these sales and agreements were concluded, Jemison owned two square miles of land, or roughly 1300 acres, and a guaranteed annual income. Her holdings had been greatly reduced since 1797, but she was able to retain the heart of her property, including her cabins and fields, and seemed satisfied with the state of affairs in 1824.

While Jemison refused to be redeemed by returning to white civilization or practicing Christianity (until, perhaps, the last months of her life), she agreed to naturalization, which is arguably the civil equivalent of ending her status as a captive. Yet Jemison's willingness to become naturalized was due to her desire to secure a measure of security for her children and their extended Seneca community—a security that would allow them to live according to their cultural traditions. Furthermore, I argue that her decision to talk to Seaver about her life was motivated in large part by a desire to retain the final parcels that she owned. As mentioned earlier in this section, Seaver's introduction notes that Jemison was "jealous of her rights" and without Thomas Clute's presence, Seaver would have been "unable to have *obtained*" her story (57; emphasis added). That Seaver was trying to "obtain" Jemison's story, rather than simply hear or record it, suggests that she was wise to be cautious. Elsewhere in the introduction, Seaver describes how Jemison "possessed an uncommon share of hospitality" and many knew of her by the time "[t]he [white] settlements increased, and the whole country around her was inhabited by a rich and respectable people, principally from New-England" (54). Clearly, Jemison was feeling pressured by the increasing isolation of her home from other Seneca. With her longstanding interest in maintaining her family and community, that no other Seneca seem to have been nearby may have precipitated Jemison's sale of the remaining land at Gardeau in 1831. To stay in a region of white settlers may have required further reacculturation that Jemison was unwilling to assent to.

Michelle Burnham makes the fascinating argument that Seaver's careful set-up of the text to show that Jemison is white, while yet reserving the role of author of the text for himself, is similar to defrauding Jemison of her land. Burnham compares *A Narrative* to the 1823 Supreme Court case of *Johnson and Graham's Lessee* v. *William M'Intosh*. In this infamous case, the Supreme Court found that Native Americans could not sell land to private citizens. Thus, Native Americans could not own land, but only be tenants upon it. The decision in *Johnson* v. *M'Intosh* recognized Native Americans as residents of the frontier only in order to create a legal mechanism for immediate removal. Burnham connects the collaborative composition of *A Narrative* with Indian Removal as a whole:

> [T]he dispossession of Native Americans was justified through a surprisingly corollary gesture: the Court acknowledged the Indians' possession of the land, but only so that the land might be legally owned by the United States. In short, the strategies of authenticating Mary Jemison as a white woman and James Seaver as author employ the same ambivalent logic of bestowal and retraction. ("However Extravagant" 329)

Burnham argues that Seaver recognizes Jemison's authority insofar as the text is written from her first-person perspective, yet he calls himself the text's author and reserves the right to assign her cultural identity. Despite the value of Burnham's comparison, I argue that we must recognize Mary Jemison's agency in that Jemison met willingly with Seaver and chose to become a US citizen in order to preserve her land and home. Far from being obligated to talk to Seaver, Jemison agreed to the interviews on her own terms. By 1823, when *Johnson v. M'Intosh* was judged, Jemison's land could not be bestowed by the federal government, much less retracted. While Jemison had first secured the Gardeau Flats as a Seneca, she had also cemented her claim to the land through naturalization as a US citizen. Similar to the adroit negotiations she conducted in 1797, Jemison remained one step ahead of the changing world around her twenty years later.

The creation of *A Narrative* can also be regarded as an unexpected success for Jemison. As other critics such as Kilcup have pointed out, a major theme of the text is Jemison's "long-term efforts to retain traditional lands" that "makes an implicit demand for environmental justice" (56). While Seaver attempts to frame her history as a conventional captivity narrative, she is able to subvert these intentions and use them to her own ends throughout the text.

Scenic and historic preservation

Mary Jemison was able to harness the genre of the captivity narrative in Seaver's written text, wisely including her ally Clute in the compositional collaboration to safeguard against fraud. I argue that much of her effort was directed at following up on her desire to remain at Gardeau Flats and to preserve a legacy for her descendents. This spectacular landscape was critical to her identity, and Seaver felt compelled to dwell upon the landscape at length, both at her urging in the life story and in a series of brief chapters that formed an appendix to the 1824 edition of *A Narrative*. In this appendix, Seaver relates various ethnographic details of Seneca life and historical information about wars, especially narratives of exploits he has gathered about Jemison's husband Hiokatoo from the unreliable George Jemison.[9] Seaver also includes a "Description of Genesee River and its Banks, from Mount Morris to the Upper Falls," in which he notes that the bluffs of the river range as high as four hundred feet (*A Narrative* [1925] 180). This description of scenic beauty includes Mary Jemison's home, which Seaver explicitly mentions. This section also claims that the Middle Falls are "not exceeded by any thing of the kind ... except the cataract at Niagara" (*A Narrative* [1925] 180). The landscape, for so many editors of *A Narrative*, is scenic in its own right, yet it is made especially significant because of Jemison's occupancy of it.

Later editors expanded these landscape descriptions. Mix added a lengthy chapter in 1842 that focused on the Genesee River. Mix links Jemison to the Genesee, pointing out that "within the space of twelve miles along that stream, she [Jemison] has since resided seventy-two years of her life ... we will give the reader a brief geographical sketch of the country" (251). Like Seaver, Mix emphasizes the scenery, writing of the area where Genishau was located: "The tract of country ... has the most delightful appearance imaginable, considering there are no lofty snow-clad peaks, deafening cataracts, or unfathomable dells, to stamp it with the appellation of romantic" (254). By employing the language of the sublime, Mix is working to show that the narrative is significant *because* of the setting. Originally, this chapter was silently added in the middle of the life story of Jemison, although later editions such as the 1925 edition moved this chapter to the appendix. Mix emphasizes the various Seneca villages that had been located on the river, and also mentions the Great Slide, noting that many trees survived the fall and were still growing decades later. Jemison's ruptured life, exemplified by her son John's violent death, which he believed the Great Slide foretold, may metaphorically be seen in a more positive light, as the "curiosity" of the site shifts to the survival and subtle rearrangement of the landscape.

Editions from the latter part of the nineteenth century and early twentieth century went even further in their depictions of the land. These later editions usually retained the earlier chapters and appendices that described the land in a narrative fashion, but also added visual representations of the landscape. Vail's 1925 edition includes more than forty illustrations, many of which are maps and photographs of the Genesee River Valley. For example, one plate is titled "Gardeau on the Genesee River: A view of Mary Jemison's flats from one of her hills" (following page 272). This photograph, reproduced in this text as Figure 2.1, consists of a foreground full of piled hay, with a horse-drawn cart on the far side of the field being loaded by two people. In the distance, the river offers light and contrast, with the impressive sandstone bluffs towering above. Another plate depicts the log cabin once occupied and owned by Jemison's daughter Betsey (following page 288). Although some of the photographs are purely scenic and do not depict human occupancy, most show how Jemison and her family lived and worked at Gardeau. The landscape is a site of labor and occupation, rather than a vacant or merely pastoral space that Jemison somehow could be construed to be not "using" to its full potential.

Figure 2.1 "Gardeau on the Genesee River: A view of Mary Jemison's flats from one of her hills." 1925.

Historical preservation was a significant motivating factor for Seaver and later editors of *A Narrative*. Seaver explains how he was enlisted to meet with Jemison and write down her story, noting that "[m]any gentlemen of respectability, felt anxious that her narrative might be laid before the public ... to preserve some historical facts which they supposed to be intimately connected with her life, and which otherwise must be lost" (54). Mix felt his geographical additions were necessary to preserve knowledge of where historical Seneca villages had been located. Yet the combination of environmental and historical preservation is perhaps best exemplified by the efforts of a later editor of *A Narrative*, William P. Letchworth, who published five editions of the text between 1877 and 1913.

Letchworth was born in New York state in 1823 and raised a Quaker. He amassed a fortune as an iron magnate and never married. He devoted many of his later years to philanthropy and progressive reform efforts, serving on the New York State Board of Charities for twenty-four years, researching and writing the important book *Care and Treatment of Epileptics*, and working indefatigably to improve conditions at prisons, orphanages, and schools.[10] In 1858, Letchworth decided to buy land along the Genesee River, a short train ride from his home in Buffalo. Notably, the only people mentioned in the geographical appendix to Seaver's 1824 edition of *A Narrative* besides Jemison were the operators of a new sawmill. More than thirty years later, the Genesee River valley had been logged over, so when Letchworth made his offer to purchase two hundred acres, most of that from a mill owner who had already logged all that was profitable, it was with an understanding that much restoration work would be necessary. Immediately upon purchasing the land, he hired the landscape architect William Webster, who described the grounds in 1861:

> Glen Iris, the country seat of Wm. P. Letchworth, Esq., is situated on the Genesee River, near Portageville, and contains about two hundred acres, finely diversified with rock, wood, and water, the scenery is picturesque and grand, and the natural advantages and capabilities of the place are well appreciated by the proprietor, and none of its natural beauties have been marred but rather improved by art. In formulating my designs, and in the execution of the work so far completed, I have strictly adhered to the natural style, and my views in all respect have been in accordance with those of the proprietor. (87)

Not only did Letchworth desire for the landscape to be maintained in a picturesque style, he immediately set about creating a literary aspect to his estate, Glen Iris. Webster notes that Letchworth converted an existing building into a library, "furnished with a large number of the leading periodicals and public journals of the day, and thrown open to the public free of charge as a reading room" (87). Letchworth belonged to a literary salon that was known as the Nameless Club. The Nameless Club consisted of perhaps ten core members, many of whom were editors and journalists for Buffalo newspapers. Two club members who eventually became quite involved with *A Narrative* were David Gray and William Clement Bryant. The liberality of the Nameless Club is suggested by the fact that two women were members (Beale 31–32). Letchworth encouraged these friends to write while visiting Glen Iris, and eventually helped publish several editions of their work in a lavish book called *Voices of the Glen*, which featured more than a dozen illustrations and gold leaf lettering on the cover of the expanded 1911 Knickerbocker Press edition.[11] Beale notes that the writers of the Nameless Club—including Letchworth's own contributions—were "unanimous in their appreciation of the place [Glen Iris]" yet the quality of their writing "range[d] widely in literary merit" (38). A glance at *Voices of the Glen* confirms this assessment. Regardless of the mixed quality of the poetry, it is remarkable that a writing retreat explicitly dedicated to the landscape was put into place nearly immediately upon Letchworth's acquisition of the estate. One of Gray's poems may serve as a good example of the poetry inspired by the scenery:

> Sweet Glen of the Rainbow, to thee there are given,
> As fresh as the day when they sprang into birth,
> All the joys and the graces we love most of earth,
> And the sunlight flings o'er thee the glories of heaven.
> So the Nameless now drink from thy pleasure-brimmed chalice,
> And pledge thee the rainbow-ideal of valleys—
> ...
> The beautiful home of a beautiful heart. (20)

Gray's poem is a conventional and stylized tribute to the Nameless Club, to Letchworth, and to the scenic Genesee River Valley and its waterfalls in particular.

As the years progressed, Letchworth continued to buy up land along the Genesee River, eventually owning roughly 1000 acres. He was fascinated by Seneca history, acquiring one of the last Iroquois Council Houses that remained standing, which he had painstakingly disassembled, moved, and then rebuilt at Glen Iris in 1872 (Beale 70–82). Two of Jemison's grandsons were present at the dedication of the building, and along with Letchworth, planted memorial black walnut seedlings that came from the tree at Jemison's grave (Beale 81). Letchworth's ongoing interest in Jemison was also manifest in his active compilation of new materials to add to future editions of *A Narrative*. Namias notes that Letchworth's research is "reflected in his private papers by an extensive correspondence with others about her life" ("Editor's Introduction" 41).

Before Letchworth could complete a new edition, he was approached with the request of returning Jemison's remains to the Genesee River valley. Her gravestone was being destroyed by souvenir-hunters, and Letchworth obliged in 1874 by providing space for the new gravesite near the Council House. He later erected a stone memorial engraved with a sentence declaring that Jemison's life was "inseparably connected with that of the valley" (Beale 89). Letchworth published a new edition of *A Narrative* in 1877, which featured additional chapters contributed by people who had known Jemison, most notably Laura Wright's "deathbed" conversion story of how Jemison allegedly began praying in her final months. While the plates for the rest of the text were identical to those used in Morgan's edition, Letchworth added wood engravings, including a landscape of Gardeau Flats, as well as many images of Seneca cultural artifacts such as earrings and tomahawks. Letchworth continued on Jemison-related projects, producing additional editions of *A Narrative*. He moved her former home, as well as her daughter's log cabin, near to the Council House (Namias, *White Captives* 161). Letchworth eventually created a museum focused on local history, with Jemison's legacy at the heart of its collections. Artifacts included a spinning wheel she had owned, as well as her will and various editions of *A Narrative* (Beale 163–65).

Yet Letchworth's experiences at Glen Iris were not always idyllic. He was forced to defend Glen Iris from private power companies, which repeatedly introduced into the state congress bills proposing dams that would have inundated Glen Iris and the Gardeau Flats, creating a fifteen-mile long lake in the Genesee River gorge (Beale 176). Letchworth lobbied furiously against these proposals, writing letters inflected by his keen interest in scenic beauty. He asked if future New Yorkers were "to be denied breathing places and the enjoyment of the natural beauty God has given us? Must everything be sacrificed to Mammon and adversely to the interests of the people?" (qtd. in Beale 177).

In the midst of these struggles, Letchworth joined a recently formed but already influential organization, the American Scenic and Historic Preservationist Society (ASHPS). The ASHPS was based in New York City, rather than Buffalo, but was clearly aligned with Letchworth's aesthetics, which understood natural scenery and historical significance to be inseparable. Now defunct, the ASHPS was most active from the 1890s through the 1920s. Its legacy is summed up as "a scattered but substantial collection of 'historic and scenic places' protected and held up as totems of civic memory—a 'memory infrastructure'" (Mason 133). Letchworth's efforts to create at Glen Iris a place that combined an appreciation for the picturesque with an interest in creating a public site for a museum, library, and historical buildings matched the ASHPS ideas of "spatializing memory" through "site specificity"; or, as Randall Mason explains, "[b]y giving historical memory lasting form in the built environment, ... the particular memory was endowed with power to reform the public at large" (140). The ASHPS also published some of Letchworth's editions of *A Narrative*. The most lavish edition of *A Narrative*, the 1925 edition by Vail, was also published by the ASHPS, as was a later, inexpensive version of *A Narrative* in 1949.

The 1949 edition omits the vast majority of ancillary materials, and in this way previews later editions such as Namias's. The ASHPS president, George McAneny, briefly introduces the volume by noting that

> [t]hose familiar with previous editions of 'The Life of Mary Jemison' will find in this edition a volume from which a series of forewords, appendices, notes and index have been omitted. This has been done

> designedly and with a view to bringing this interesting and unique work to the attention of a wider public. … To scholars it is enough to say that these may be found in full in previous editions.

The scope of Letchworth's involvement with the ASHPS went well beyond his work on a few editions of *A Narrative*, however. As private power interests continued to threaten Glen Iris and Gardeau with proposals to erect dams, he worked to create a greater degree of security for the land and for the legacy of both himself and Jemison. He met with the ASHPS to discuss how to will to the state of New York Glen Iris and his 1000 acres that stretched for three miles along the Genesee River, with the condition that the ASHPS would retain permanent custody of the land. In 1907, the state accepted Letchworth's and the ASHPS's offer, calling the land "Letchworth Park" (Beale 181–84). The Glen Iris estate and the Gardeau Flats are at the center of what is now a fourteen-mile long park that is about one-mile wide along the river valley.

One of Letchworth's final acts was to commission a memorial statue of Mary Jemison from an accomplished sculptor, Henry Kirke Bush-Brown, who also was a trustee of the ASHPS. Bush-Brown memorialized Jemison at Letchworth State Park in 1910 with a large bronze statue depicting her striding forward with her child on a baby board; see Figure 2.2 for a photograph of the statue. Namias notes that Letchworth and Bush-Brown corresponded for four years about this statue and that much research went into creating as authentic a likeness as possible. Bush-Brown also consulted with Arthur C. Parker, who had a long and distinguished career as an archeologist. Parker was a descendent of many important Seneca leaders. As part of the lengthy document that records the context and speeches given at the statue's dedication, Parker is quoted as saying, "Your Mary Jemison is one of the most accurate, if not the most accurate, studies of New York ethnology which I have ever seen. It is not only a monument to the heroic captive but a fitting memorial and an accurate one to the nation which adopted her" (American Scenic and Historic Preservation Society, 235). Parker, along with two women who were descendants of Jemison (Thomas Kennedy née Sarah Jemison, and Carlenia Bennett), actually unveiled the statue (253).

Not only is the statue historically accurate, it retains a vibrancy today that is unusual for memorials created in the early twentieth century. Namias writes that "[t]o Bush-Brown's credit, he has Jemison moving … The bronze statue projects a strength which Jemison undoubtedly had, and it is physical as well as moral" (*White Captives* 163). Some parallels exist between Bush-Brown's 1910 statue of Jemison and Benjamin Victor's 2005 statue of the Paiute woman and author Sarah Winnemucca, which is in the National Statuary Hall in Washington, D.C., and represents the state of Nevada.[12] Both statues convey a sense of movement, with intricate fringed dresses and long hair flipped in front of the depicted women's shoulders, which seem to sway. That such a comparison can be made is a testament to Bush-Brown's vision, as Victor's statue is considerably more dynamic than nearly all other statues in the National Statuary Hall—many of which were crafted by contemporaries of Bush-Brown. The memorial statue remains at what is now Letchworth State Park, and a likeness is prominently featured in gold-leaf on the cover of the 1925 edition of *A Narrative*. A photograph of the statue is also reproduced as a frontispiece.

Figure 2.2 "Statue of Mary Jemison at Letchworth Park." Frontispiece, 1925.

Today, the Genesee River Valley is known as the "Grand Canyon of the East." Its dramatic waterfalls likely would have been inundated if not for Letchworth's efforts, which were inspired by his belief that the landscape had a unique and site-specific importance created through the combination of scenic value and historical value. Much of this history was derived from Mary Jemison's legacy, and in particular, from her collaborative text with James Seaver, *A Narrative of the Life of Mrs. Mary Jemison*. While she could not have foreseen this in her final years at Buffalo Creek, her many descendants have a very clear legacy that has been preserved at Letchworth State Park. The landscape thus remains very much linked to Jemison's extra-textual legacy. As mentioned earlier, Jemison concludes *A Narrative* by listing her large family and how they remain connected to the Genesee River. Jemison is "not just a figure of history," but rather is a well-known ancestor to many people in New York state and a "very real figure affirming the possibility that whites and Indians might have lived together peacefully" (Namias, "Editor's Introduction" 43). Without Jemison's collaboration with Seaver on the surprisingly subversive *A Narrative*, the Great Slide might well be at the bottom of a reservoir.

Notes

1. Jemison's biographical details are fairly consistent, though June Namias points out that the ship the Jemison family was on, the *William and Mary*, sailed from Belfast several times in 1742 and 1743 and it is thus uncertain which voyage the growing family was part of and which year Jemison was born.
2. As each edition published added materials to the text, I refer back to these editions regularly in this chapter, and Appendix 1 is very helpful while tracking these changes.
3. Unless otherwise noted, page citations refer to the 1992 University of Nebraska Press edition of James E. Seaver's *A Narrative of the Life of Mrs. Mary Jemison*, edited by June Namias.
4. Reverend Asher Wright began his missionary work with the Seneca at Buffalo Creek Reservation in 1831 and married Laura in 1833. While the letter by Laura Wright seems embellished and less than fully trustworthy—appearing more than forty years after the events recollected—the Wrights should not be seen as exploitative missionaries. Both Asher and Laura rapidly became fluent in Seneca, and Asher published books on Seneca culture and language. Asher also denounced the 1838 removal of the Seneca from Buffalo Creek in an 1840 letter published by the Society of Friends entitled "The Case of the Seneca Indians in the State of New York Illustrated by Facts." Thomas Abler's 1992 article on Wright in *American Indian Quarterly* forthrightly argues that "not all missionaries carry extreme ethnocentrism as part of their personal cultural baggage" (25).
5. I disagree slightly with Kolodny, as *A Narrative* was popular upon its publication in 1824 and was able to both compete with and draw upon the existing mythology of Daniel Boone first promoted in John Filson's "The Adventures of Colonel Daniel Boon" (1784). Nonetheless, in the twenty-five years since

The Land before Her was published, *A Narrative* has undergone a resurgence of interest, which may account for my own view of the text's influence. After all, Jemison's renewed popularity is due in part to Kolodny's important work in showing how women reshaped male fantasies of a wilderness Eden.

6. This name was later given to a stream that feeds into the Genesee at Letchworth State Park by Letchworth; like many other small tributaries in the area, it features dramatic waterfalls.
7. Interestingly, both of Jemison's husbands had names that mirrored their homes in some way—or perhaps conflated the Native American with the land. Sheninjee married Jemison while she lives at Shenanjee. Hiokatoo was apparently called "Gardow" by many locals, and although Jemison notes that "[i]t has been said that … my land derived its name from him," that common belief was "a mistake" (121). She explains that the place known as Gardeau, alternatively spelled Gardow, instead "derived its name … from a hill that is within its limits, which is called in the Seneca language Kautam" (121). While these links between married life and land are striking, the full nature of these connections remains uncertain.
8. The one place Jemison specifies general problems faced in her life from the Seneca is on the final page of *A Narrative*, where she admits that "some of our people … [thought] I was a great witch" and that some of her children had lighter skin "which used to make some say that I stole them … I never thought that any one really believed that I was guilty of adultery" (160). By and large, however, Jemison seems to have enjoyed a supportive relationship with most Seneca.
9. Namias omits the appendix, choosing to reprint the 1824 edition without the ancillary materials in her 1992 University of Oklahoma Press edition. She states that this is because "the aim of this edition is to recover Mary Jemison's own words as fully as possible" ("Editor's Introduction" 44). Along with the addition of Namias's detailed and helpful introduction, this omission marks the two primary differences between her edition and the 1990 Syracuse University Press edition.
10. Irene Beale's 1982 biography of Letchworth, subtitled *A Man for Others*, is the most recent book-length overview of his life. The biography is scholarly, though its documentation of sources is incomplete.
11. The text was first published in 1876 but was expanded and reissued as: *Voices of the Glen*. Ed. Henry R. Howland. Administrator, William Pryor Letchworth. New York: Knickerbocker Press, 1911.
12. For more on the legislature, cultural, and creative processes that led to Benjamin Victor's Winnemucca statue, see Kyhl Lyndgaard's article "Sarah Goes to Washington: How Sarah Winnemucca Entered the National Statuary Hall." *Nevada Historical Society Quarterly* (2015).

3 Scientific and sympathetic collaboration

Edwin James and John Tanner

The life of Edwin James (1797–1861) is bookended by the Lewis and Clark Expedition (1803–06) and the Civil War (1861–65). James was born in Vermont and graduated from Middlebury College in 1816. He rapidly became acquainted with a circle of prominent scientists and presented early botanical and geological work at the Troy Lyceum in New York (Benson, "Edwin James" 7, 14–25).[1] The 1819–20 expedition to the Rocky Mountains led by Major Stephen H. Long employed James as a botanist and surgeon. When James is mentioned today, it is usually in connection with the first recorded ascent of Pikes Peak, which occurred during the Long expedition. In fact, Pikes Peak was briefly known as James Peak in his honor. James's solid botanical descriptions and specimens are recognized as offering the first recorded examples of many North American plants, especially those collected in the Front Range of the Rocky Mountains. One was a blue columbine (James called it *Aquilegia caerulea*), which is now the state flower of Colorado. Another was an alpine shrub commonly known as cliffbush or waxflower, which was given the scientific name *Jamesia americana* Torr. & Gray (Williams 22–23).

After the expedition, James never returned east for lengthy periods, choosing instead to work on the frontier in what are now Minnesota, Wisconsin, and Michigan. In 1830, James published an unusual captivity narrative in collaboration with the "White Indian" John Tanner, with whom he worked in the Michigan Territory. James's work on this narrative included a striking and timely critique of Indian relocation and remains noteworthy for its fidelity to Tanner's life story. James eventually settled near Burlington, Iowa, a small city on the Mississippi River, and his farmhouse became a noted stop on the Underground Railroad.

Despite such a rich and productive life, however, James's name has fallen from the covers of both his major works. *Account of an Expedition from Pittsburgh to the Rocky Mountains* (1823) and *A Narrative of the Captivity and Adventures of John Tanner* (1830), as well as many of his articles, were published with his name listed as editor or compiler rather than as author. The captivity narrative James wrote in collaboration with John Tanner, an illiterate man, was recently republished by Penguin with James's name expurgated from the book.[2] James's work engaged key national concerns of western exploration, natural history, Native American relocation, and slavery. His principled stands for preservation of lands and animals in the Trans-Mississippi West and his opposition to Indian relocation are remarkable today, yet his legacy does not fit neatly into established literary or historical categories. While the collaborative nature of James's writing has obscured his popular legacy, his willingness to collaborate demonstrates the integrity he displayed during a period of intense scientific and literary competition. Due to the uniqueness of James's voice and perspective, this chapter focuses as much on his work as editor as upon Tanner, the captive with whom he collaborated.

First, I situate James's collaborative relationships and argue for the importance and uniqueness of his literary voice. Next, I examine James's early career as botanist for the 1819–20 Long expedition to the Rocky Mountains in order to show his progressive belief that the Great Plains—or "Great American Desert"—should be preserved through game laws, and even be reserved for Native American hunters. In the heart of this chapter, I discuss James's work on Tanner's captivity narrative and explain how this cross-cultural, collaborative text resisted Andrew Jackson's 1830 Indian Relocation Act. Finally, I examine the ramifications of the personal and political disputes that occurred in the 1830s between James, Tanner, and important Indian agent and ethnologist Henry Rowe Schoolcraft.

Much of the recent scholarship on Mary Jemison and Black Hawk has attempted to trace their voices and presence in texts produced by unreliable editors. Arnold Krupat's influential book *For Those Who Come After: A Study of Native American Autobiography* is partly responsible for this. Black Hawk provides Krupat's example of a Native American autobiographer—an assessment with which other critics continue to grapple, as neither Black Hawk nor Jemison is likely to have had a conception of the genre of autobiography. Antoine LeClaire and J. B. Patterson, the translator and editor of *Life of Black Hawk*, and James Seaver, the editor of *A Narrative of the Life of Mary Jemison*, furthermore, are not seen as historically important figures, which may have contributed to scholars' eagerness to discern their editorial presence in order to release the authorial voices supposedly held captive within the texts. James, on the other hand, was well-connected to many famous explorers, politicians, and scientists. His goal in publishing Tanner's narrative was not primarily commercial, but rather was scholarly and motivated by his opposition to national Indian policy. Yet, unlike the case with Black Hawk's and Jemison's editors, James's work as editor with John Tanner has rarely been discussed.

Perhaps James has been elided because his authority and point of view have not fit with the early goals of theorizing American Indian autobiography, which focused on legitimizing and restoring the legacies of early Native American authors. Only recently have these authors been recognized as important voices in early American literature, so this chapter complicates Native American autobiography by looking at the white collaborator just as much as the author, arguing that the full context of the cultural production cannot be divorced from the text.

Gordon Sayre's article "Abridging between Two Worlds" does suggest that Tanner's narrative is "torn between two genres and two motives of potential readers" (483). According to Sayre, readers find that the narrative alternates between two modes. The first mode is that of an authoritative captivity narrative, full of ethnographic detail and information, but the second is that of an American Indian autobiography, as Tanner fails to make a successful return to white culture in the end. Sayre gives an accurate sketch of James's career over the course of two paragraphs; however, the bulk of his essay is focused solely on Tanner, as his argument looks at how the book can be fruitfully read as Native American autobiography. Elsewhere, Sayre unequivocally states that if Tanner had been of mixed race, "the book would be identified as a nineteenth-century Native American autobiography, and would by now be published in several editions and be the topic of many academic studies" ("John Tanner, Métis" 136).

Collaborative partnerships between a white man and a member of an ethnic minority are now often broadly seen in retrospect as necessarily exploitative. Therefore, while his progressive attitude and body of work suggests he was ahead of his time, James has been overlooked once again. That James is very much a silent collaborator throughout the narrative (exclusive of the introduction and appendices) allows readers to focus solely on Tanner's role in producing the narrative. This editorial choice is made possible by the quality and nature of the collaboration, which defies conventional tropes of captivity.

By collaborating with both white and Native authors, James managed to shed many of the individual prejudices that Krupat identifies in writers of the American Renaissance (*Ethnocriticism* 52). James's efforts to engage honestly with the people and places he encountered on the frontier resulted in some extraordinary passages. In his introduction to Tanner's narrative, for example, James calls out "worthless squatters" who want Native Americans removed from the frontier; he states that these people, as well as many politicians, "consider the region west of the Mississippi as a kind of fairy land where men can feed on moonbeams, or at all events, that the Indians, when thoroughly swept into that land of salt mountains and horned frogs, will be too remote to give us any more trouble" (xxxii). James here critiques the romanticization of the West and the idea of the Indian as noble or tragic.

James's most productive years were during the 1820s and 1830s, when he addressed the twin questions of westward expansion and Native American relocation. My prologue, which discusses the moccasin flower, helps illustrate the prevailing attitude of white Americans during the relocation era. James's work is notable for its unusually forceful and well-informed resistance to exploitative land use policies and environmental attitudes. His work illuminates the historical relationship between ethnicity and landscape. Native Americans were systematically removed from the frontier landscape during the early nineteenth century: "The effort, both literary and political, of removing Indians into a romanticized, mythic 'elsewhere,' far to the west in the place of the setting sun, produced a pervasive image of 'romantic Indians'" (Faery 191). This socially constructed and self-serving image of Native Americans depicts them as happily giving way in the face of white settlement, and even facilitating the process through self-sacrifice. Rebecca Blevins Faery discusses several texts that perpetuated this literary treatment of Native Americans, such as *Hobomok* and *Hope Leslie*, but she struggles to find dissenting voices—in part because people like James, who collaborated with others, did not produce works that honor the idea of an individual creative genius. As daring as James's ideas were, he backed them up with cautious, rational arguments. With multiple names on the title pages, but the names of Long and Tanner also in the title, it is perhaps not surprising that James, who did the bulk of the writing, is not always identified strongly with his own books. A closer look at James and his career illuminates his struggle against better-known narratives that celebrated, or at least accepted as inevitable, manifest destiny and the Indian Relocation Act during the early nineteenth century. James's writing is much more engaged than that of his contemporaries because of his impressive and consistent efforts to see the native people and landscapes of the frontier on their own terms.

Despite James's striking arguments, well-supported by evidence and written in an engaging style, his uncommon viewpoints have drawn persistent criticism that has been often personal in nature. For example, in a 2003 book that borrows James's own words for its title, *'A Region of Astonishing Beauty': The Botanical Exploration of the Rocky Mountains*, author Roger L. Williams dismisses James's social concerns by saying that "the obvious merit of such causes [as Indian welfare, the temperance movement, and abolitionism] has never protected them from being attractive to cranks" (29).[3] The dismissal of James's importance due to his challenges to the status quo is, I believe, an unfortunate mistake. Clearly, James's literary achievements remain important: they are topical in the sense of being of contemporary interest and also in the sense of being grounded in place, or topos. The collaborative nature of these texts highlights the integrity of James's argument, even as collaboration served James's own career and ambition less well. Despite significant success, competition from other leading writers and scientists, especially Thomas Nuttall and Henry Rowe Schoolcraft, may have discouraged him from writing in his later years.

James's local knowledge of the people and places he encountered in both the Great Plains and in Michigan Territory, and his concerns for larger national debates and government policy, caused his writing to challenge uncritical proponents of westward expansion. By calling attention back to James, we can better understand the intense national debates that surrounded the expansion of the US during the antebellum period. James's resistance to this expansion was much more original and intense than what is seen in any other major exploration texts, and in most captivity narratives. While other texts share in the project of disseminating information about the western frontier, they almost never question the underlying assertion that the nation should expand westward without regard to the consequences of expansion for native peoples and the nonhuman environment. James's use of a captivity narrative not to sensationalize events, but rather to convey scientific and ethnographic information, is a mark of his conscientious resistance toward the status quo of the 1820s and 1830s.

Expedition to the Rocky mountains: A preservationist view

Throughout *Account of an Expedition from Pittsburgh to the Rocky Mountains*, James displays a willingness to grant autonomy and worth not only to Native Americans, but also to the plant and animal communities of the Great Plains. In addition to his major collaborative volumes, James worked closely with leading scientists in the eastern United States, such as John Torrey and Amos Eaton.

Despite some early successes, James was unable to support himself financially with his scientific work upon graduating from college. James's agony at his lack of gainful employment ended in 1819, when he joined the expedition headed by Major Long to the Rocky Mountains.[4] A bundle of 90 letters to his older brother John, recovered in 1983 from a philatelist, shows James's concerns regarding his career (Miles 35).[5] Along with scattered letters in other collections, the letters to John James offer a rare glimpse into James's personal aspirations and fears in his early twenties, uninflected by the moderating tone for the national audience that his official expedition report required. While his later works are replete with descriptive passages that aspire to scientific precision, the letters, not surprisingly, are more daring and emotional. Because Edwin James was intensely private in his later years, he instructed his housekeeper to destroy all of his letters and papers shortly after the death of his wife (Benson, "Edwin James" 334). Thus, letters from James are more readily available than letters sent to him.

A 1991 analysis of the extant letters by Carlo Rotella notes how James's

> letters to John have a timeless quality, expressing the self-doubt and premature cynicism of an aspiring young professional anxious about his career and dignity [... The letters are] characterized by despair shot through with wild surges of hope, betraying shame at his economic and emotional dependence on his family (28).

Rotella overemphasizes James's disagreements with Long during the trip, focusing on the complaints that James sent to his brother without noting that the collaborative relationship with Long was repaid with loyalty and a correspondence that James maintained for the rest of his life. While James was undeniably frustrated by the limitations of the expedition that were imposed by Long—due in large part to budget restrictions—the two do not seem to have had any major confrontations, and they worked together closely on the written account of the expedition upon their return. Long complimented James's writing style in an 1855 letter to James, calling his prose style "a feast of Reason and flow of Soul" (qtd. in Benson, "Edwin James" 335). This phrase was in service of an attempt to convince James to work on a multi-volume work about Long's career, to be called *Mississippiania*, and definitively shows that the two men had a long and complementary relationship.

Nonetheless, James's letters to his brother confirm his frustrations with Long's leadership, which stemmed from the expedition's minimal funding from the federal government: "[o]ur scientific and enthusiastic commander encamped on a plain of sand at the distance of 24 miles from the base of the mountain and informed me that he allowed me three days to make what examination I wished among them" (qtd. in Rotella 30). These three days were the occasion of James's unprecedented climb to the summit of what is known today as Pikes Peak—the adventure for which James is best remembered. James writes in his journal "[i]t is not without a feeling of regret that we reflect that our only visit to these 'Palaces of Nature' is now at an end. Our opportunities for seeing and admiring these grand and beautifull scenes have been few and are now ended" (July 19, 1820). A similar passage in the published account confirms that James's journal was mined extensively for publication: "[i]t was not without a feeling of regret, that we found our long contemplated visit to these grand and interesting objects, was now at an end" (2:50). James's words, so personal in his letters and journal, remain highly descriptive in the expedition's official account. James's approach can be discerned in a statement made at the beginning of his book's lengthy appendices, which he writes are no longer "compatible with the humble style of a diary, which we thought convenient to retain" (2:330).

In the process of adapting his own journals to the more formal and scientific prose required of an expedition narrative, James makes an effort to keep a personal voice in the narrative sections of the text. This resulted in a highly readable and widely accessible text, more literary than Schoolcraft's 1820 *Narrative Journals of Travels from Detroit*, for example, which too often reads like a diary that records little more than the weather, and often does so in incomplete sentences. On the other hand, when describing the pleasures of finding wild grapevines, James wrote the following sentence, which is both complex and personable:

> We indulged ourselves to excess, if excess could be committed in the use of such delicious and salutary fruit, and invited by the cleanness of the sand, and a refreshing shade, we threw ourselves down, and slept away with unusual zest, a few of the hours of a summer afternoon (2:126–27).

Indeed, the quality of the prose was noted a decade later in the first issue of the journal *The Western Monthly Magazine, and Literary Journal*. In a brief essay entitled "American Literature," the writer argues that while exploration narratives in Britain are widely read and appreciated, American exploration narratives are "coldly received, ... among the heaviest lumber of the bookstores" (7). However, the essay continues, "public sentiment is wrong" and "[t]he accounts of both of Col. Long's expeditions are well written, and parts of them composed with great elegance; the adventures are novel and curious, and the scientific facts collected, numerous and useful" (7). While not a bestseller, *Expedition to the Rocky Mountains* is recognized as a particularly well-written and valuable American exploration narrative.

Notwithstanding the important discoveries made by the Long expedition and the fine prose in which the expedition narrative is written, the group suffered serious criticism that is emblematic of James's entire career. Their journey, according to many critics from 1820 to the present, was severely misinformed. Not only did Long's crew follow the wrong river for weeks, they also labeled the Great Plains as the "Great American Desert" in both their maps and text. James was not the first to attach the word "desert" to the Great Plains, as he demonstrated by citing this usage from texts by Thomas Nuttall and Miguel Venegas (2:386–87). The Long expedition's maps and descriptions of the Great American Desert were blamed by many politicians for having delayed westward expansion for decades. The sentiments of the expedition's leadership are clearly stated in the narrative: "[t]he traveller who shall, at any time, have traversed its desolate sands will, we think, join us in the wish that this region may forever remain the unmolested haunt of the native hunter, the bison, the prairie wolf, and the marmot" (James, *Account of an Expedition* 2:161). Their underlying argument was that the area would serve as a natural impediment to expansion, allowing the young country to consolidate its population in the East while buffering its territory from British encroachment. Only a few years after the Treaty of Ghent ended the War of 1812, strategic national defense interests were at play here, in addition to the growing sympathy James had for Native American rights. A final argument against settling this region came directly from the grueling experience of crossing the southern Great Plains during the height of a dry summer.

Few of the plants James collected in his role as botanist were described in the expedition's account, which was published in 1822, before the taxonomic classification of specimens could be fully completed. James was torn between his willingness to collaborate and his desire to receive due credit for the discovery of new species. In a December 1820 letter to Eaton, James complained that Nuttall had "taken many of those [new plants] which should have been mine out of my hands. He will [...] be out with a second edition [...] before I shall have an opportunity to say any thing" (qtd. in Benson, "Edwin James" 116). James's concerns about competition at the beginning of his career were understandable. To allay these worries, an initial catalogue of plants James compiled was read in Philadelphia at the American Philosophical Society in 1821 (Benson, "Edwin James" 116–18). After this public reading of his findings, James was sufficiently confident to turn over his specimens and notes to Torrey, who presented three papers on James's discoveries in the mid-1820s to the New York Lyceum of Natural History. Torrey explained that James, "whose zeal in prosecuting his favourite science is so well known [...] has kindly permitted me to commence the publication of the discoveries he made" (Torrey 30). Those papers were not only the first reports on western alpine flora, they were also the first accounts to be published in America that follow the natural system of plant classification, rather than the Linnean system (Ewan 14–15). Because Torrey carefully credited James for his discoveries, this partnership offers an excellent example of how nineteenth-century scientific collaboration could work.

Compiling the narrative of the expedition was also a successful collaborative experience for James. Only twenty-six when the volumes were published, James felt fortunate to have been given the opportunity by Long, who was thirteen years older and could easily have reserved that honor for himself as commander of the expedition. James, Long, and naturalist Thomas Say worked together closely, although only James was paid to work on the project full-time (Benson, "Edwin James" 89–95). James was well aware of their assistance and of his unique opportunity to compile the account. Even on the first page, James gives credit to Say and Long: "[t]hose gentlemen have indeed been constantly attentive to the work, both to the preparation of the manuscript and its revision for the press" (1:1). The artist Titian Ramsey Peale, son of Philadelphia naturalist and museum curator Charles Willson Peale, was also a member of the expedition and remained in contact with James after their return. James's personal journal even features sketches by Titian Ramsey Peale, including a macabre drawing of a human skull found near the Platte River on June 18, 1820. Almost exactly two years later, a journal entry records that Peale accompanied James on an "excursion" to the New Jersey coast together (June 14, 1822).[6]

Expedition to the Rocky Mountains has proven to be a widely influential book, for many reasons. Some commentators have traced the book's influence on novelists, especially James Fenimore Cooper and Washington Irving.[7] The passages that deal with landscape descriptions, bison, and Native Americans, especially the Pawnee, were immediately recognized for their literary and documentary value, and were thus mined by novelists as well as celebrated by reviewers. For example, a lengthy 1823 review in *The North American Review* was so complimentary as to call the expedition "better qualified and fitted out [...] than the company of their distinguished predecessors, Lewis and Clarke [sic]" (Anon. 242). One example of an episode that Easterners found so fascinating was a "dog dance" performed by a group of Konzas (Kansas):

> Each warrior had, besides his arms, and rattles made of strings of deer's hoofs, some part of the intestines of an animal inflated, and inclosing a few small stones, which produced a sounds like pebbles in a gourd shell. After dancing round the fire for some time, without appearing to notice the strangers, they departed, raising the same wolfish howl, with which they had entered (1:135).

To highlight the significance and import of this scene, an etching by Samuel Seymour of the ceremony was also included with the narrative. As in many of the images from the narrative, a few select members of the expedition are depicted. In this case, three white men are seen to the left in the far background, and James may well be one of the three. Rather than dominate the image or alternatively be removed entirely, the group is a rather subtle and deferential component of the etching. Indeed, they appear to be shorter than the Konzas, and do not imply domination or superiority in any way. This shared attitude is a good example of James's overall vision of his work as compiler of the narrative and member of the expedition. Rather than erasing him entirely from the background, however, I believe readers of his books and viewers of this etching should remember his willingness to depict Native Americans and the environment as objectively and respectfully as possible.

After the Louisiana Purchase, the desire to learn more about Native Americans in the West became a national concern, and one that James in his later career would fully embrace. In the expedition to the Rocky Mountains, Say had primary responsibility for conducting Native American studies. Native Americans were then popularly considered a fit subject for the study of natural history, and were thus not placed in the same category as the white explorers. While considering Pawnees as specimens to be taxonomized and described as if they were flora or fauna is understandably problematic for modern readers, the Long expedition's unusually careful and respectful attitude regarding Native Americans—and megafauna such as bison, grizzly bears, and wolves—stands in stark contrast to the attitudes of most nineteenth-century explorers.

While describing attempts by Peale to trap wolves, for example, James writes that "[t]he wonderful intelligence of this animal, is well worthy of note," and he asserts that the ability of the wolf to circumvent traps in order to gain the bait safely is the "result of a faculty beyond mere instinct" (1:172). Over the course of eight pages dedicated to describing the varieties of wolves he encountered, James never uses a pejorative adjective or connotation. Bears are also described in mostly appreciative terms, with the significant exception of wounded bears that charge hunters. James dutifully relates encounters hunters have had with bears, but it is his own encounter that carries the most emphasis by strength of its elaboration and detail:

> [a] half grown specimen [...] came to me, and rearing up, placed his paws upon my breast; wishing to rid myself of so rough a playfellow, I turned him around, upon which he ran down to the bank of the river [...] and swam about for some time (2:55–56).[8]

This surprisingly benign and humorous passage about a captive bear describes James's experiences at the Missouri Fur Company in Cape Girardeau. A line drawing by Titian Ramsay Peale depicts these bears in a similarly playful way, and is reproduced in this text as Figure 3.1. Rather than emphasizing the bravado of hunting bears or the dangers faced on such an expedition, this drawing suggests the natural curiosity of bear cubs.[9]

Figure 3.1 "Three Bears," Cape Girardeau, Missouri, Long Expedition. Titian Ramsay Peale, 1820. T. R. Peale Sketches (B P31.15d, #123), APSimg5690. Courtesy of the American Philosophical Society.

James further argues for game laws that would halt the indiscriminate slaughter of bison already commenced by white hunters, and he notes repeatedly that he and the other leaders of the expedition ordered their hunters to refrain from killing animals when they already had meat. For example, on June 19, 1820, the expedition was forced to halt travelling westward toward what is today the Nebraska-Colorado border in order to hunt for food. After sighting an immense herd of bison, James wrote, "as we had already killed a deer, and were supplied with meat enough for the day, none of the party were allowed to go in pursuit of them" (1:469). James recounts only a week later that "it was with difficulty [that] we restrained our hunters from slaughtering many more [bison] than we needed" (1:481).[10] This ethical stance is highly unusual in literature from the early nineteenth century, in which authors generally boast of their participation in hunting sprees, and even give bison hunting as a primary reason they are on the expedition. For example, Washington Irving, in *A Tour on the Prairies*, writes that his party shot eighteen turkeys in a single grove of trees, but "[i]n the height of the carnage, word was brought that there were four buffaloes in a neighboring meadow. The turkeys were now abandoned" (143–44). After killing three bulls, Irving notes that the meat of these bison was "meagre and hard, at this season of the year. A fat buck yielded us more savory meat for our evening's repast" (145). Obviously, shooting only as many animals as needed was not a concern for Irving, as it was for James.

Irving focuses on characters and impressions, leaving aside any qualitative judgments on the nature of the small expedition he joins. Rather, he crows repeatedly about how dangerous his adventure is, writing that "the purport of the following pages [is] to narrate a month's excursion to these noted hunting grounds, through a tract of country which had not as yet been explored by white men" (11). Irving figures the western prairies as a place to prove himself. James, on the other hand, sets up the expedition as a purely scientific inquiry. Rather than hunting, for example, James and the others are guided by the goal of attending to "the phenomena of nature, to the varied and beautiful productions of animal and vegetable life" (1:1).

Avoiding bravado at every turn, James's prose occasionally turns toward whimsy:

> We found a constant source of amusement in observing the unsightly figure, the cumbrous gait, and impolitic movements of the bison; we were often delighted by the beauty and fleetness of the antelope, and the social comfort and neatness of the prairie dog (1:474).

Furthermore, James never loses sight of the ramifications of the westward migration of white settlers. He quotes various colonial writers who describe bison in Virginia and Kentucky, asserting that there "can be no doubt" of this animal's former historical range. James places blame squarely upon American expansion: "[t]his process of extirpation has not since been relaxed, and the bison is now driven beyond the lakes, the Illinois, and southern portion of the Mississippi rivers" (1:472).

Indeed, the slaughter of bison was a deliberate attempt by some whites to destroy the Native American cultures that depended upon the animal, and so the Long expedition was therefore advocating preservation not only of game, but of the Native Americans who depend upon game for survival. James concludes that

> It would be highly desirable, that some law for the preservation of game, might be extended to, and rigidly enforced in the country, where

> the bison is still met with: that the wanton destruction of these valuable animals, by the white hunters, might be checked or prevented (1:482).

The next writer to make so forceful a preservationist argument was George Catlin, eighteen years later, when he famously proposed a national park to preserve both Native American hunters and buffalo. Catlin linked the two in romanticized language: "Of man and beast, no nobler specimens than the Indian and the buffalo—joint and original tenants of the soil, and fugitives together from the approach of civilized man" (254). Catlin's position, both proto-preservationist and anti-relocation, is one that James first formulated, and from which James never retreated.

John Hausdoerffer argues provocatively that Catlin "preferred not to consent to the cultural and environmental domination he saw and resented around him," yet that "what we prefer and what we practice when submerged in the ideological abyss of a historical moment are often quite different things" (19). This argument—which I find compelling—suggests that while Catlin remains an admirable figure who objected to Native American removal, he tailored his work to appeal to the masses, and in this sense may have inadvertently supported Indian relocation. James did not labor under the same commercial constraints as did Catlin. While not wealthy, James always had another job by which to support himself. Indeed, while collaborating with John Tanner on a translation of the New Testament into Ojibwe, James wrote, "I sometimes regret that I get only experience for my pains, but even experience is worth the trouble I take." (qtd. in Pammel 183–84).[11]

James's objections to mainstream American views about the frontier only hardened as he became better acquainted with individual Native Americans. In the decade following the Long expedition to the Rocky Mountains, James traveled extensively between the East and the frontier in what is today Minnesota, Wisconsin, and Michigan. His second major publication demonstrates a more focused argument regarding ethnicity and the environment, and his career took a remarkable turn as James engaged fully with the political climate of his time and place. By moving to the captivity narrative, James shifted away from the restrictive mode of government compiler and toward a popular genre that allowed—and even expected—passionate and melodramatic prose.

Edwin James and the unredeemed captive

Published early in 1830, *The Captivity and Adventures of John Tanner* depicts the fragmentation of Native American life as the major fur trading companies introduced guns and liquor on a large scale along the waterways of the Great Lakes and beyond. John Tanner's early childhood was spent with his white pioneer family in the backwoods of Kentucky, near Elk Horn and later near the mouth of the Great Miami River along the Ohio-Kentucky border. After being kidnapped at age nine in 1790 by a small group of Ojibwe who hoped Tanner might be a replacement for the deceased son of a man named Manito-o-geezhik, Tanner underwent a strenuous northward march for hundreds of miles that greatly reduced the chances of escape or rescue.[12] After two miserable years with the family of Manito-o-geezhik, Tanner was adopted by an Ojibwe/Ottawa woman. Tanner became almost completely assimilated into his new family's culture. Tanner showed great affection and respect for his adoptive mother, Net-no-kwa, admiring that "[s]he could accomplish whatever she pleased, either with the traders or the Indians; probably, in some measure, because she never attempted to do any thing which was not right and just" (27). Tanner spent nearly thirty years living as a subsistence hunter and trapper in the prairies and forests of what is today Minnesota, North Dakota, Ontario, and Manitoba before contacting his biological family in 1819 and 1820 in Kentucky and Missouri. Because Tanner was culturally Ojibwe/Ottawa at that time, unable to speak much English or even sleep indoors, the reunion was very short-lived, and Tanner soon returned to the Michigan Territory. James met Tanner in 1827 at Fort Mackinac, a fort on what is now a Michigan island in Lake Huron, where Tanner was working as an interpreter and James was working as an army surgeon (Benson, "Edwin James" 172).

In response to President Andrew Jackson's proposed Indian Relocation Act of 1830, James wrote that "the project of congregating the Indians ... not only *west of the Mississippi*, but ... in those burning deserts which skirt the eastern base of the Rocky Mountains, is, perhaps, more pregnant with injustice and cruelty to these people than any other" ("Introduction" xxxi).[13] James was a vocal and forceful critic of Indian relocation policies—a criticism made more authoritative by his first-hand experiences in the Great Plains. While his counterproposals varied, all were motivated by a sense of social justice and informed by his expansive knowledge of the physical environment that supported Native American cultures. James suggests that the ideal policy, though impossible in practice, may be to "*let them alone*" (xxxii; emphasis in original). James holds out no hope that a single removal will be sufficient to implement government policies of relocation: "let every Indian be removed beyond the Mississippi, how soon will the phrase be changed, to *west of the Rocky Mountains?*" (xxxii). His argument, of course, proved to be accurate, and suggests that some people of the time knew that this ever-westward displacement of Native Americans was the trajectory of Indian removal policy.

The popular genre of the captivity narrative had long been used to justify the killing and displacement of Native Americans, starting at least as early as Mary Rowlandson's *The Sovereignty and Goodness of God* (1682). Yet the collaborative work of James and Tanner in 1830 tried to do just the opposite. *A Narrative of the Captivity and Adventures of John Tanner (U.S. Interpreter at the Saut de Ste. Marie) During Thirty Years Residence Among the Indians in the Interior of North America* was intended to offer a more accurate representation of Native American life, working against contemporary views that romanticized a "doomed" race—views popularly expressed in novels such as James Fenimore Cooper's *The Last of the Mohicans* (1826). Most literary texts from the early nineteenth century, including Cooper's, follow what Gerald Vizenor has called "manifest manners:" the "notions and misnomers that are read as authentic and sustained as representations of Native American Indians" (5).

Rather than depicting manifest manners, however, Tanner's narrative stands as a highly authentic text. Two years before it was published, a brief article in the *Christian Watchman* gave advance notice by quoting from a letter to their editor claiming that Tanner,

> [b]eing a man of a strong, unbending, and naturally descriminating [sic] mind, he has been converted by his editor into one of the purest and most fruitful channels, through which information respecting the manners, mode of thinking, &c. of the red men, has flowed into our sea of literature (163).

Interestingly, the word "converted" here suggests the conversion narrative, a genre that parallels and at times overlaps with the captivity narrative, yet in the text Tanner is skeptical of both Native American prophets and Christian missionaries. Furthermore, less cautious editors of texts such as Jemison's narrative add Christian elements in later editions, in part to sell more copies. To a degree, notices such as the one just cited can be described as conventional publicity, but the tone that is struck does stand out as distinct from notices that described other texts in more sensational terms. Tanner's forthcoming text, it is concluded, is valuable for the quality of its information rather than the sensationalism of its narrative.

Tanner's narrative was published in 1830, after a delay caused by the difficulties the printer G. & C. & H. Carvill had with the many Native American words James insisted on retaining for accuracy (Benson, "Schoolcraft" 316–17; Fierst, "Return" 27). Nonetheless, the book was published before passage of the Indian Removal Act. The 1830 Tanner narrative consisted of three parts: an introduction, by Edwin James; "Part I," which is Tanner's life story, told in first-person; and, "Part II," which is about half the length of Part I. Part II is James's ethnographic and philological work, some of which cites John Tanner as its source. The 1830 book reproduces a portrait of Tanner by Henry Inman that was painted during Tanner's visit to New York with the lithograph prepared by Cephas G. Childs (Lee 388; Fierst, "Strange Eloquence" 227–28). In this portrait, Tanner's hair is cropped, his skin is quite light, and he is wearing a formal coat with a high collar (Figure 3.2). This is the only image of Tanner in existence, and it has been reprinted in many different editions. Tanner, this visual image seems to suggest, has been fully converted to white civilization.

Figure 3.2 John Tanner, frontispiece from the 1830 edition of Tanner's narrative. Courtesy of the Clarke Historical Library.

The idea of Tanner as a "white Indian" bears further examination, as the term was in vogue during the 1820s. A white Indian, after all, was one of the only possible formulations in the early nineteenth century where skin color and cultural identity were divorced from one another and repackaged in a new way. The story of Tanner fascinated the public well before the publication of Tanner's narrative in 1830. The 1801 journal of a fur trader named Daniel W. Harmon (published in 1820) contained a note on Tanner: "he speaks no other language except theirs. [...] He dislikes to hear people speak to him respecting his white relations, and in every respect but color he resembles the savages with whom he resides" (80). Harmon thus suggests that race is only one marker of identity.

A very different account, published in 1819 and notably entitled "The Indian Captive Reclaimed," glosses over the fact that Tanner cannot speak English and needs an interpreter in order to communicate with his brother. The anonymous author dwells instead on Tanner's physical resemblance to his brother Edward, going on to note that John "has nothing of the savage wildness and ferocity in his appearance [...] and is remarkably clean in his person; of robust, manly appearance; temperate habits; drinks no ardent spirits" (127). Some readers may have desired that this story match their expectations for how an Indian captivity narrative should end; namely, with Tanner's reintegration into his biological family and their church. This was not to be, even though some later editions, such as those edited by James Macauley and Louise Erdrich, do not offer accurate information on Tanner's life after his family reunion. Tanner did in fact live for several decades after reuniting with his brother—approximately the same length of time as his original "captivity"—but in his later life, Tanner was somewhat of a recluse on the edges of Sault Ste. Marie. One local man reported that "'Tanner was a regular Injun; more of an Injun than any of the Injuns, and a d-----d mean Injun too'" (qtd. in Steere 254). This folksy characterization suggests another potential formulation of a white Indian, although this man in general "did not take an optimistic view on the Indian question" (Steere 254). White Indians are expected to embody Indian stereotypes rigidly if they want to pass as Indian.

In Tanner's narrative, the difficulty of determining Tanner's identity is on display. For example, while Tanner is in the midst of a delicate and potentially dangerous attempt to recover his horse from a rival tribe, he is confronted by an Assiniboine who calls him an Ojibwe. Tanner replies that he is "not altogether an Ojibbeway" (138). Tanner's identity, as curious and fascinating as it was to white Americans at the time, also marked him as unusual among the Native Americans with whom he lived. Thus, Tanner himself may be suggesting that he is neither fully Indian nor fully white. Less nuanced portrayals of Tanner as mixed race also have been published. While Walter O'Meara's *The Last Portage* (1962) is a well-written and sympathetic biography of Tanner, it unfortunately features a dust jacket upon which Tanner is depicted with a face half-red and half-white, "as though he had fallen asleep under a sun lamp" (Fierst, "Strange Eloquence" 237).

Part II of Tanner's narrative is wholly James's work, and consists of several chapters of vocabularies and catalogues, not unlike the appendices in *Account of an Expedition from Pittsburgh to the Rocky Mountains*. The catalogue of plants, which is arranged by subheadings such as "Ever greens, or cone bearing trees," is several pages long, covers approximately 150 plants, and lists the Ojibwe word for each plant alongside the common English name while occasionally offering notes to further explain etymology and/or medicinal uses (293–99). Tanner is credited in Part II as a source, alongside many others James was studying and consulting at the time. In the third chapter of the appendix, entitled "Music and Poetry of the Indians," many different songs are given which were compiled from a variety of sources. Notably, all the sources are given equal weight, regardless of their cultural background or standing in society at the time. "*Song of Chi-ah-ba, a celebrated Ojibbeway Medicine man, at the administration of his remedies*" begins on page 376, while only few pages later appears "*Song of the warriors about to start on a war party,*" which ends with a note saying "[t]his song has been published, and illustrated, by Mr. Schoolcraft" (381). Henry Rowe Schoolcraft is recognized in James's work—but with no more weight than Native American informants.

While James conducted impressive scholarship and held strong opinions against relocation, James did have some limits to his sympathy toward Native American culture, as seen in his reliance on common tropes regarding the state of Native American literature. In "Music and Poetry of the Indians," James comes across as dismissive of Native American literary production at times, suggesting that this is a "barren field, offering little to excite industry, or to reward inquiry" (334). Despite his disclaimer, James writes nearly fifty additional pages in this chapter, which suggests that he found the subject less "barren" than his formulaic opening of the chapter suggests. Those introductory pages of the chapter, which were probably composed a bit earlier than the rest of Tanner's narrative, were drawn from the notes he compiled throughout the 1820s. Therefore, language that comes across as oddly dismissive often represents preconceived notions and scholarly summaries of other information James was familiar with, rather than his rapidly growing first-hand knowledge of the subject.

Tanner's skepticism regarding the visions and superstitions of the Native Americans is also interesting from a collaborative perspective.[14] At times, Tanner uses various prophetic hunting strategies to locate game. James must have been worried about the reception of such episodes among white readers, as he tries to downplay this content in the introduction, writing that "[m]any will find their confidence in [Tanner] much impaired, when he tells of prophetic dreams, and of the fulfilment of indications, and promises, necessarily implying the interference of invisible and spiritual beings" (xix). As a careful scientist, James consciously followed up sensational episodes with rational explanations. When Tanner's adoptive mother tells him she had a vision of where to kill a bear, for example, Tanner claims that it is because she tracked the bear "into a little thicket, and then circled it, to see that he had not gone out" (48). Tanner explains further that "[a]rtifices of this kind, to make her people believe she had intercourse with the Great Spirit, were, I think, repeatedly assayed by her" (48). Other examples of this scientific and rationalist skepticism are easy to find, and include an unintentionally humorous story in which Tanner attempts to kill an otter with his bare hands to disprove the belief that otters cannot be killed without a weapon. "I beat him, and kicked him, and jumped upon him, but all to no purpose. I tried to strangle him with my hands [...]" (208). The otter, caught on a frozen lake, is pummeled for more than an hour. The experiment concludes with the line, "I was at last compelled to acknowledge that I was not able to kill him without arms" (208–09). Obviously this is not the diction of a man whose English is poor. Tanner also displays great skepticism toward prophetic movements, a belief that may have been emphasized by James, who was involved with a local Baptist mission.

The kinds of cross-cultural collaboration that produced captivity narratives in the early nineteenth century were often dependent upon the proximity of the literate editor to the residence of the former captive, and indeed that is how James began recording Tanner's life story. The James-Tanner collaboration began in 1827, when James and Tanner began working on Tanner's narrative at Fort Mackinac. Rather than seeking out Tanner, James had been assigned by the US Army to his posts at Fort Mackinac and Fort Brady in the Michigan Territory—the places where Tanner, coincidentally, found employment. Nonetheless, James may appear as early as 1820 in Tanner's narrative. The text notes that in 1820, near Cape Girardeau, Missouri, Tanner "saw some of the gentlemen of Major Long's party, then on their return from the Rocky Mountains" (257). This sentence is not expanded upon. Fierst suggests the two men had met in 1820 and were aware of each other ("Strange Eloquence" 231). It seems more likely to me that they did not meet for another seven years. While in Cape Girardeau, James was very sick, spending a month in recuperation while the rest of the expedition straggled to their respective homes further east. Tanner had the opposite problem: after sleeping indoors at his new-found relatives' homes, he became very ill, and rallied only after returning to his habit of sleeping outside.

In an additional coincidence, Long's next expedition met Tanner in 1823 along what is now the border between Minnesota and Ontario. Long does not mention in his journal having previously met Tanner, saying only that "his history is an eventful one, sketches of which have been circulated thro' the medium of the Newspapers of the U. States" (214). James, however, was not on that expedition. While Long had written to James offering him a position, the letter failed to reach him in time. In the meantime, William Keating was hired provisionally, with the understanding that if James could catch up, Keating would forfeit his pay and continue service on a volunteer basis (Kane et al 24). Keating's expressed hopes to publish a more comprehensive version of Tanner's life story were frustrated by Tanner's slow recovery from a gunshot wound, but Keating noted that "Tanner had promised to supply us with the particulars of his life and adventures, and with a full account of the manners and habits of the Ottawas and Chippewas" (128). Tanner was shot during an attempted move of his Native American wife and their children to Mackinac Island, where he could work as an interpreter. Apparently his wife was unhappy with the plan, and went so far as to convince another man to attempt to murder Tanner and run away with her. Tanner's life was marked by violence and conflict, and this incident is all too emblematic of his struggle.

James, frustrated by being replaced by Keating, redeemed himself by recording the story of John Tanner in 1827. So in the competition to produce new stories and information, Keating may be said to have been scooped by James. James had spent the previous three years posted at Fort Crawford in Prairie du Chien (in present-day Wisconsin, on the Mississippi River). During this time, he had grown increasingly interested in Native Americans, and had even delivered three studies on Native American languages and customs to the Philadelphia and New York libraries in 1826 (Benson, "Edwin James" 197–98). James was sent to Fort Mackinac in 1827, before being sent to Fort Brady at Sault Ste. Marie the next year (Benson, "Edwin James" 172). The collaboration between James and Tanner during that year at Fort Mackinac was extraordinarily productive. Tanner delivered the manuscript of the captivity narrative to a publisher in New York the following summer, only to find upon his return that his job at Mackinac had been terminated in his absence. So Tanner headed to Sault Ste. Marie, where he was hired by the local Indian agent, who was none other than Henry Rowe Schoolcraft. This move also marked a reunification with James, who had since moved to Sault Ste. Marie to work at Fort Brady.

Despite the fact that Tanner was illiterate and even had to relearn English during the 1820s, James's name and introduction have been completely purged from recent editions of the narrative, one of which is part of the well-regarded Penguin Nature Library. The collaboration with Tanner was so seamless that Penguin repackaged the book as an autobiography, despite leaving James's editorial footnotes in the text, and despite several scenes in the book that clearly show Tanner's inability to speak English during much of his life. James's name appears neither in the book's front matter nor in Louise Erdrich's introduction, while James's 1830 introduction and the second part of the original book have been excised. Those deleted sections consisted of James's ethnographic and philological work, some of which cited Tanner as a source. Impressively, James's text is scrupulously documented with sources.

The resulting text that Penguin has reprinted regularly through the 1990s and 2000s is an unusual book. Rather than being a captivity narrative emphasizing Tanner's return, the latest reprint now functions as an American Indian autobiography, as Gordon Sayre argues convincingly ("Abridging" 480–99). The title on the cover of the book has been changed from *A Narrative of the Captivity and Adventures of John Tanner*, which capitalized on the widely read genre of the captivity narrative, to *The Falcon*. The new title emphasizes the "Indianness" of Tanner's identity, and indeed several of the book jackets for these Penguin editions underline this message. Tanner's name, as given in the text, is indeed Shaw-shaw-wa Be-na-se, or falcon. The translation of this name is given as "the falcon" under the portrait of Tanner, as well as on page 15. However, in the original 1830 appendix, no such bird is listed, and the closest match is "Shaw-shaw-wa-ne-bais-sa—Swallow" (307).[15] Regardless, this discrepancy was also in the 1830 frontispiece. The cover of the 1994 edition is dominated by a large feather. The 2000 edition features a painting of a man with a peace pipe, elaborate silver jewelry and a bison skull. This painting is by Albert Ferris, who was a member of the Turtle Mountain Band of Chippewa, and suggests a fully transculturated autobiographer; its title is "Prayer over Buffalo Skull Antler." The most recent edition's cover art, however, features a version of Inman's portrait of Tanner. This choice may represent a step away from "Indianness," and closer toward James's and Tanner's authorial intentions, though it certainly continues to emphasize the idea of this text as an autobiography.

This removal of James is not unique to the Penguin editions. While some European versions of the narrative appeared shortly after the first edition in 1830, Philadelphia's Lippincott, the same publisher who had published Schoolcraft's memoirs, published the second American edition of Tanner's narrative in 1883.[16] This volume was edited by James Macaulay, who is advertised on the title page as author of the book *All True*. James merits only a passing reference in Macaulay's brief introduction, which explains the reasons for not publishing James's introduction and Part II: "The book [...] contained [...] disquisitions and essays upon the natural history, traditions, languages, religions, and the political relations of the various tribes [...] to the American Government [...] Much of the matter is of limited, and some of it of temporary and bygone importance" (vi). Macaulay's editorial choices argue that the Native Americans' culture and survival was of less importance to him than a lively adventure story that would appeal to "youthful readers, both of England and America" (vi). By changing the title to *Grey Hawk*, Macaulay significantly alters the text by de-emphasizing the captivity narrative genre. Macaulay's cover art includes a peace pipe and teepee, while the main image is of a man in leather hunting clothes firing a gun at a bear. Rather than a transculturated member of a Native American community, this depiction is suggestive of a lone white fur trader in the tradition of a figure like Jedediah Smith. Furthermore, the figure is depicted in a landscape that more closely resembles an alpine area in the Rocky Mountains than the relatively flat land of northwest Minnesota and southern Manitoba, where Tanner spent much of his life.

Macaulay also alters the written text, injecting romantic and hyperbolic language that often jars the reader from James's understated style. In James's text, the narrative begins by stating: "[t]he earliest event of my life, which I distinctly remember, is the death of my mother. This happened when I was two years old, and many of the attending circumstances made so deep an impression, that they are still fresh in my memory" (1). Macaulay rewrites these two sentences as follows:

> The first event of my life of which I have any recollection is the death of my mother. This happened before I was three years old. I have no distinct remembrance of her person, but only of the love with which

> she loved me, and of the aching void in my young life when I saw her no more, and heard her voice no more. (1)

From this melodramatic and partly fictionalized opening, Macaulay's remaining changes are similar in tone: he adds stock emotions that he thinks will appeal to a young adult audience. James, on the other hand, wrote the original text alongside serious linguistic and ethnographic work that he hoped would be a significant contribution in his field. Macaulay finishes the story with a sort of epilogue that claims nothing adventurous happens to "interpreters" and so "Grey Hawk" had uneventful later years, and of "the date or circumstances of his death, we have been unable to learn" (301). Macaulay then discusses "Henry L. [sic] Schoolcraft" for several paragraphs in glowing terms. The book ends with two shorter pieces, called "The Sun Dance of the Sioux Indians," and "The Hudson's Bay Fur-Hunters," neither of which are connected to Tanner's narrative except in the most superficial and tangential way.

While Tanner certainly had become transculturated, the text itself is so reliant upon the skills of James as an editor that James's contributions must not be overlooked. Retrieving James's unique contributions and goals allows us to understand the narrative within the richness of its contextual production, rather than in isolation from the important events that led to its publication. Despite the accuracy of the text, Tanner's narrative may fail as Native American autobiography if it were published, as originally intended, with James's lengthy 1830 introduction arguing against the Indian Relocation Act, or with the original ethnographic and linguistic section that is nearly half as long as the narrative itself. The unconventional aspects of the text are precisely what make it noteworthy, yet editors such as Macauley and Erdrich conventionalized the text in various ways to conform to the expectations of audiences from the 1880s to the present day.

By any measure, James's efforts in recording Tanner's story resulted in a remarkably accurate portrayal of the life of a subsistence Ojibwe or Ottawa hunter. Tanner's narrative has even been used by scholars in other fields as source material for their research—a contribution rarely made by literary texts. Susan Sharrock, for example, uses passages from Tanner's narrative as historical evidence of "polyethnic coresidence units" among the Cree and Assiniboine (111). That any additions to Tanner's story by the literate collaborator James are nearly impossible to ascertain is to James's credit: other texts, such as James Seaver's *A Narrative of the Life of Mrs. Mary Jemison* (1824) and Henry Rowe Schoolcraft's *Algic Researches* (1839), have been widely criticized for corrupting and disguising the oral narratives upon which they are based.[17]

Yet even the supposed authenticity that has allowed Tanner's *Narrative* to be repackaged as Native American autobiography has been criticized. June Namias finds the story unconvincing purely because of the extensive recollections of hunting: "[Tanner] told of every bear and elk he killed and probably a good number he never saw" (*White Captives* 78). Unlike Namias, I believe that these lengthy and repetitive stories should instead be taken as evidence of Tanner's assimilation into an oral culture, and James's allegiance to presenting the story accurately.[18]

Critics also found the story difficult to fit into their notions of what a captivity narrative should consist of. One 1830 reviewer of Tanner's narrative, who was extremely positive about the book, found his impression of Native Americans quite different from the unapologetically stark narrative Tanner and James present:

> the book ... presents as painful a view of the Indian character as ever fell under our notice; we know no work of equal extent that is so replete with the story of drunken brawls, of low artifices, of petty thievings, of idle vaunts, and cowardly desertions.
>
> (Anon., "A Narrative" 123).

The British review in *The Albion* is primarily made up of extended passages from Tanner's narrative that emphasize severe privation and cruelty. The best example of this is when a young girl accidentally burns down their shelter in the winter, and Tanner punishes her by forcing her to sleep in the snow, away from the warm ashes (67). James addresses this episode in his introduction, saying that the "Chippewyans" have different social and cultural practices, and "actions considered among us not only reprehensible, but highly criminal, are among them accounted shining virtues" (xxi). The book is thus uniquely poised to dispel romanticized notions of Native Americans, while yet asserting distinct cultural practices. The mostly successful struggles of Tanner to avoid alcohol and to subsist upon the land are fully developed and show an individualized and human effort to live a traditional Ojibwe/Ottawa life in the face of white encroachment. Even later writers, such as Catlin, who are often characterized as sympathetic to Native Americans, are ultimately unable to avoid romanticizing or essentializing their subject; therefore, James's efforts rank among the very best representations of the period.

While subtle, James's presence in the text may be discerned in how the names of plants, animals, and places are given in both English and Native languages with remarkable specificity. Here is one straightforward example from Tanner's *Narrative*: "Then we put our canoes into a small stream, which they called Begwionusk, from the begwionusk, or cow parsley, which grows upon it" (45). Cross-cultural collaboration is notable here, as no one language or culture is privileged over the other. This sensitive treatment also extends to the people that appear in the text. As Fierst notes, powerful white figures such as Major Long or Lewis Cass, the Governor of Michigan Territory, are mentioned only in passing, while the most fully developed characters are Native Americans, including Tanner's adoptive mother, Net-no-kwa, and others like Pe-shau-ba and Little Clam ("Strange Eloquence" 223–25).

James's presence is easy to remove from the narrative precisely because he kept his promise that "[Tanner's] whole story was given as it stands, without hints, suggestions, leading questions, or advice of any kind, other than 'to conceal nothing'" (James, "Introduction" xix). Unlike so many other writers, who offered versions of this formulaic promise of authenticity and failed to honor their pledge, James's primary concealment was of his own hand in the text. The collaborative relationship between James and Tanner was in fact a great success. The quality and accuracy of descriptions of the natural world, as well as of the native inhabitants of the frontier, give Tanner's narrative great value to a range of readerships. At the time of its 1830 publication, however, the ramifications of the publication were likely an unpleasant surprise.

Sault Ste. Marie: Removals and collaborations of the 1830s

"In an attempt to aid this unfortunate individual in addressing his countrymen, it seemed desirable to give his narrative, as nearly as possible, in his own words" (James, "Introduction" xviii). James thus opens *A Narrative of the Captivity and Adventures of John Tanner* by suggesting that Tanner has been slandered and deserves a chance to tell his side of the story. One function of captivity narratives is to clear the name of the captives and dispel rumors about them. This type of appeal was fairly common to captivity narratives, as captives were often viewed with suspicion, especially if they had been with the "enemy" for an extended period. Indeed, even Mary Rowlandson's narrative is introduced by an authority figure, almost certainly Increase Mather, who wrote "I hope by this time none will cast any reflection upon this Gentlewoman" (66). Thus, James is momentarily turning to what was already a 150-year-old convention to cast his book as a captivity narrative. Captivity narratives, after all, were one of the only publishing venues open to people such as Tanner, who would otherwise have been unable to gain power or influence due to their race, gender, class, or educational standing.

Also, the meaning of "countrymen" seems to have shifted: when the book was published, James was referring to whites, yet the new introduction by Erdrich—which replaces James's introduction—suggests that Tanner's "countrymen" were Native Americans. She writes that this book is "one of the very few in the captivity genre that appeals strongly to Native Americans. But then John Tanner was culturally an Ojibwe, and as such he is claimed by many to this day" (xi). Erdrich thus points out the appeal of Tanner to a readership that was virtually nonexistent at the time of its original publication. With the exception of such figures as Jane Johnston Schoolcraft, who is covered later in this chapter, very few Native Americans were afforded an education on the frontier of the 1820s and 1830s.

James's name was removed from his literary efforts in large part because of his willingness to collaborate with and give credit to those partners who worked with him. Starting in 1828 and continuing into the early 1830s,

Tanner, James, and Henry Rowe Schoolcraft all lived in the small outpost of Sault Ste. Marie, on the shores of Lake Superior. Schoolcraft had originally hired Tanner as an interpreter, and James worked there as a surgeon for the US Army. Tanner's position would seem to have been a coup for him, as Schoolcraft was then rapidly gaining national prominence. Schoolcraft's national debut, much like James's, came as a writer of an account of an expedition published in the early 1820s. Unlike James, however, Schoolcraft hurriedly published the work without the help or even the knowledge of other expedition members. David Bates Douglass, another member of the expedition, was led by Schoolcraft to believe that they were to collaborate, and was thus unpleasantly surprised by Schoolcraft's publication of the expedition narrative mere months after the expedition had returned (Jackman and Freeman xvii–xxi; Bremer 45).[19]

Schoolcraft soon turned the focus of his prodigious output to Native American myths and languages, publishing several widely read and highly influential books. His early Native American publications are considered the inspiration for Henry Wadsworth Longfellow's fictionalized representations of Native Americans in his best-selling long poem *The Song of Hiawatha* (1855). Yet in addition to his early publication of *Narrative Journals of Travels from Detroit*, which upset Douglass, Schoolcraft's work as an ethnologist rarely acknowledges his specific debts to his collaborative partners. Despite this, his work continues to receive acclaim. William Clements, for example, argues that

> the way in which Schoolcraft presented the narratives resulted in what for him was an *accurate* version of materials that were basically *literary*. Not to have made changes in them would have betrayed their aesthetic quality and made them less literary (183; emphasis in original).

Clements thus justifies Schoolcraft's methodology, considering him a folklorist and popularizer in the tradition of the Brothers Grimm. James, on the other hand, claims that the only changes he made to the narrative Tanner told him were to "retrench or altogether to omit many details of hunting adventures, of travelling, and other events [...] It is probable the narrator might have proved more acceptable to many of his readers had this retrenchment been carried on to a greater extent" ("Introduction" xix). While the commercial success of Schoolcraft over James is indisputable, so too is the ethnographical accuracy of James's text, which is widely considered superior to most of Schoolcraft's work. Arguably, James's stated willingness to include much information about Tanner's life as a subsistence hunter is valuable not only because of the accuracy and detail entailed, but also because retaining such stories is a sign of his respect for both Tanner and the Ojibwe and Ottawa cultures to which he belonged.

One of Schoolcraft's major collaborative partners was his wife, Jane Johnston Schoolcraft. Of Ojibwe descent, she was an excellent writer who had full access to Native American folklore and other stories as a result of her family connections and her fluency in French, English, and Ojibwe. Henry Rowe Schoolcraft, aided by this silent and unusual collaboration with his talented wife, quickly eclipsed James's career. Jane Johnston Schoolcraft's complete writings were for the first time been published under her name in *The Sound the Stars Make Rushing through the Sky: The Writings of Jane Johnston Schoolcraft*, edited by Robert Dale Parker (2006). Parker's introduction offers background on the community of Sault Ste. Marie and describes the extraordinary influence Henry Rowe Schoolcraft held in it as Indian agent and husband to Jane Johnston. The Schoolcrafts' collaborative relationship, at times harmonious and at other times less so, is described by Parker in more detail and in a more even-handed manner than in earlier scholarship on the subject. Michael T. Marsden, for example, suggests that that Jane's death and funeral—both of which Henry missed while in Europe—are emblematic of a loveless marriage of convenience in which Jane was "content to play to the role of helpmate" (162). Parker's intensive textual editing suggests a mostly positive collaborative life for the Schoolcrafts, especially prior to the death of their son in 1827. Parker also publishes many of her poems and letters, which had previously been unknown or available only in archival collections. Still, he remains puzzled at Jane's lack of publications, noting that although her "choice not to publish remains mysterious, it is not so mysterious as Emily Dickinson's refusal to publish a generation later, because Dickinson had an accessible publishing infrastructure that Schoolcraft never had" (36). Furthermore, Parker considers Jane Johnston Schoolcraft's writing to be more literary than her husband's, though of course Henry is better known for his prolific output than for his prose style (36).

Laura Mielke considers the Schoolcrafts' work as an example of "familial collaboration," pointing out that it is unknown to what extent "Henry's use of Jane's compositions signals a marriage in which the wife readily submitted to the husband's control in the name of Christian morality" (140).[20] Regardless, Jane Schoolcraft is cited as a writer only in a local publication they produced from December 1826 to April 1827, called *The Literary Voyager* or, alternatively, *The Muzziniegun* (Mielke 141; Parker 34–36). In the 1839 text that perhaps stands as the most influential book Henry Schoolcraft produced, *Algic Researches*, a dozen individuals are listed by name; the first two listed are "Mrs. Henry R. Schoolcraft, Mr. William Johnston, of Mackinac" (1:57, note). William Johnston was Jane's brother. Considering all evidence available, I believe that their early partnership—in marriage, literary collaboration, and as prominent members of the leading family of the community—was more positive than not, and it also seems clear that their relationship deteriorated over time. The marriage and literary collaboration ended in 1842 with Jane's death.

Ten years earlier, James's departure from Sault Ste. Marie in 1832 marked the end of five years of collaboration with Tanner. James's translation of the New Testament into Ojibwe, entitled *Kekitchemanitomenahn gahbemahjeinnunk Jesus Christ, otoashke wawweendummahgawin*, clearly identifies Tanner as co-author. James also published various short papers on Native American languages during this period. These ethnographic and linguistic studies are marked by clear respect for their subject. For example, James's address to the American Lyceum in New York, delivered in 1833, contains many appreciative passages such as the following: "Instead of the easy word *horse*, he [Ojibwe speakers] employs, in accordance with the combining and explanatory genius of his own language, the compound *babashekokashe*, a word worthy of Linnaeus himself, signifying *the animal that has a single nail on each leg*" ("Essay on the Chippewa Language" 443). This later address mirrors James's early writing about Native American languages:

> The free and easy spirit of the Indian is carried even into their language, and may be recognized there, by its absolute destitution of a single word, drawn from the language of a civilized people [...] [Native Americans] universally and in every instance reject the names which they originally hear for such men and things, and apply others, which they readily invent.
>
> (*Account of an Expedition* 1:343)

James uses the word "civilized" not as a marker of superiority, but instead to allow for the coexistence of multiple languages. James summarized his attitude in 1828 by noting that he had "said enough" to show his "leaning towards that class of inquirers, who are disposed to admire the flexibility and compass of the Indian languages, rather than those who despise their poverty" ("Review of a Grammar" 401). While James did not argue that Native American place names, for example, were more appropriate than English language names, he did find that languages such as Ojibwe could describe the natural, phenomenal world more exactly than could English—no small matter for a scientist. He notes that "in their adjectives [can be observed] a kind of preciseness, which ours often want. In the quality *sharpness*, they designate by one adjective that kind which belongs to a point, by another that of an edge" (401). James's admiration of Native American languages was one of many reasons he believed that the American government should treat Native American tribes with respect.

Schoolcraft's staunch support of the Indian Removal Act is apparent in a letter he wrote to Governor Lewis Cass in late 1829: "The great question of the removal of the Indian is [...] put to rest [...] It only requires the moral courage necessary to avow the principle, and to reconcile the moral feelings of the friends of the Indians" (qtd. in Bremer 189). While the letter does not explicitly single out James, Schoolcraft did later engage James more directly following the publication of Tanner's narrative.

James and Schoolcraft waged various letter-writing campaigns against each other, some of which involved Tanner. For example, a December 9, 1831, letter James wrote to Schoolcraft asked why Tanner had not been given back pay for some months in 1829 when Tanner had been suspended as interpreter (Benson, "Schoolcraft" 321). Schoolcraft's reply came four days later. Schoolcraft angrily replied to James that

> instead of employing your Knowledge of letters and habits of reflection, in explaining and pointing out to [Tanner] the leading truths which oppose his success in civilized society, you leave it to be inferred that you are not unwilling to condescend to flatter error, when that error promises to minister to your personal gratifications.
>
> (qtd. in Benson, "Schoolcraft" 321–22)

In the end, Benson concludes—very fairly toward Schoolcraft—that "it appears Schoolcraft had no wish to treat Tanner harshly" (323). What is certain is that by 1831 James and Schoolcraft were openly feuding in this small frontier outpost. Their points of conflict included funding for a Native American school, and a battle over Schoolcraft's tacit permission to allow alcohol in the settlement over the objections of the local temperance society and despite laws to the contrary (Fierst, "Return to 'Civilization'" 23–36; Bremer 128; Benson, "Schoolcraft" 320–24).

The greater problem was not one of professional competition, but rather arose from the fact that Tanner was caught between employer and collaborator, and, being relatively powerless, personally suffered more than did James or Schoolcraft. Perhaps Tanner's most keenly felt blow occurred when a law was pushed through the territorial legislature granting power to the local sheriff to "remove Martha Tanner, daughter of John Tanner" and place her in "some missionary establishment, or such other place of safety as he may deem expedient" (Territory of Michigan). For John Tanner, 1830 was a tumultuous year. First, at the age of fifty, he saw his life story published with the help of Edwin James. President Andrew Jackson's Indian Removal Act was, after spirited debate, passed by Congress and signed by Jackson on May 28. Then, this extraordinary local law, which named Tanner and his daughter Martha individually, was passed on July 30, 1830 (Territory of Michigan), while Tanner was removed from his post by Schoolcraft on August 12 (Fierst, "Return to 'Civilization'" 30). Tanner's version of the story appears in a remarkable letter he dictated to another daughter in 1837 and sent to President Van Buren. The letter is worth quoting at length:

> I take opertunity [sic] this day to reach my words to you with tears calling upon you for help. because of my long Sufferings by the hand of Mr. Henry Schoolcraft. It is 7 years past since he lays his hands upon me. Governor Cass placed me hear to be an Iterpreter [sic] for Government. and Mr. henry Schoolcraft took the office away from me on purpose to give it to his Brother in law George Johnson And He took my Daughter away from me also which was keeping house for me and stript me alone and throw me down to the dust.
>
> (qtd. in Fierst, "Return to 'Civilization'" 25)

This law echoes the national policy on the level of the tiny frontier settlement of Sault Ste. Marie. As for Martha Tanner, she eventually became a teacher in Michigan Indian schools (Steere 247). While it is possible that Martha had been mistreated by her father, other evidence suggests that John Tanner was not a pariah in 1830. Tanner, who had not been told where his daughter had been taken, searched for her, soon reaching Governor Cass. Cass wrote Schoolcraft personally to urge Tanner's reinstatement, writing that "I really pity him very much. He seems to me a forlorn heart broken man" (qtd. in Fierst, "Return to 'Civilization'" 31). Furthermore, Tanner remained for some years as a reasonably productive member of the community, working for both James and the Baptist missionary Abel Bingham.

James and Schoolcraft did not cite each other's work after Tanner's narrative was published in 1830, although James did cite Schoolcraft in the appendices to that text. In his authority as Indian agent at Sault Ste. Marie, Schoolcraft terminated James's freelance services as medical doctor in 1830; however, James remained employed at nearby Fort Brady (Benson, "Edwin James" 183). One of the most famous linguists of the era, Peter Duponceau, found fault regarding the way Indian languages were handled by Schoolcraft in the 1834 publication of *Narrative of an Expedition through the Upper Mississippi to Itasca Lake*. Duponceau wrote to James, "It seems to me that [Schoolcraft] wishes to occupy the whole ground alone. He has named none of his predecessors, either Heckewelder, Pickering, yourself, or any other but condemned them generally" (qtd. in Bremer 238). Schoolcraft may have been jealous of Tanner's allegiance to James. Richard Bremer's biography of Schoolcraft makes clear that James's next project, a collaboration with Tanner on a translation of the New Testament, "might well put Schoolcraft's own reputation as a Chippewa scholar in the shade" (128).

This translation is reminiscent of John Eliot's 1663 "Indian Bible."[21] Eliot's efforts were very highly regarded in progressive circles in the 1820s and 1830s, and James was well aware of the parallel, writing in Tanner's narrative that Eliot's Bible was the best model available for "adapt[ing] to the use of the Ottawwas and Ojibbeways, in the country about the lakes" (422). Indeed, the final pages of Part II of Tanner's narrative consist of a translation of the first chapter of Genesis into Ojibwe.

Interestingly, a Narragansett Indian man in the Massachusetts Bay Colony, known alternately as James Printer or James the Printer, who is mentioned in Rowlandson's narrative, was an integral typesetter for both the 1663 edition as well as the 1685 edition of Eliot's Bible. Eliot wrote an 1683 letter to Robert Boyle praising James Printer, noting that "we have but *one man*, viz. the *Indian Printer*, that is able to compose the sheets and correct the press with understanding" (qtd. in S. Drake 57). In this way, Printer may have served a collaborative role in the production of Eliot's Bible, which at the least would have been much less accurate without his work as typesetter—and less likely to be praised by somebody like Edwin James more than 150 years later. James Printer served Samuel Green as an apprentice for fully sixteen years, leaving only at the outbreak of King Philip's

War—a departure that by the Puritans was seen as an illegal and immoral breaking of his contract (S. Drake 56–57). Kathryn Zabelle Derounian, Jill Lepore, and Neal Salisbury note that James Printer—himself captive during the war by the same group that held Rowlandson—was involved with Rowlandson's ransom negotiations and almost certainly helped typeset some early editions of her book (245; 145–49; 49). Eventually given amnesty by the white colonists, James returned to the printshop until as late as 1709, when his own name appeared on the title page of a bilingual psalter: "Boston, N.E. *Upprinthomunne au* B. Green, & J. Printer, *wutche guhtiantamwe Chapanukke ut New England*" (qtd. in S. Drake 57). James Printer, who worked on both editions of Eliot's Bible with King Philip's War in between, is a fascinating bilingual, literate figure whose life suggests cross-cultural collaboration began with the importation of the very first printing presses to North America.[22]

Despite the inspiration that Eliot's Bible—and perhaps even James Printer—afforded to Edwin James, at least one unfavorable review of the James/Tanner New Testament was published: Wm. Herr, a Michigan missionary, wrote that James "does not understand the language, and was merely assisted by John Tanner, who but partially understands the English; and consequently, the translation is such as might be expected from such hands" (10). Perhaps not surprisingly, Schoolcraft's review was also negative, stating that "it would be more than reasonable to expect that a person [Tanner]" who was

> nurtured in the depths of Indian prejudices and superstitions, should succeed in catching the true meanings, far less the spirit [...] of this book. And it cannot excite surprise that the translation is often so wide of the true meaning, as to render the book worthless (*Information* 536).

Despite such commentary, James persisted in working on Ojibwe language books for years, and many of his editions were used in frontier churches and schools during the middle of the nineteenth century.

After James left Sault Ste. Marie, the remainder of Tanner's life degenerated to a near-subsistence existence on the periphery of the settlement, and the only information for this period regarding Tanner comes from a couple of arrests for killing livestock (Fierst, "Return to 'Civilization'" 36). Tanner disappeared in 1846 after being publicly blamed for the murder of Schoolcraft's brother—a crime he almost certainly did not commit.[23] When Schoolcraft publicly discussed James and Tanner in his 1851 memoirs, his feelings remained quite strong. Schoolcraft claimed that Tanner felt betrayed by James's ability to leave the settlement: "Dr. James [...] made a mere pack-horse of Indian opinions of him, did not suspect his fidelity, and put many things in his narrative which made the whites about St. Mary's [Sault Ste. Marie] call him an old liar" (601).[24] Schoolcraft's memoirs make clear that Tanner, to whom Schoolcraft refers as "Caliban," was not forgiven either (343). Regardless of their personal feelings, the three men were at the epicenter of a national debate, even while living in a tiny frontier settlement. Although Henry Rowe Schoolcraft continued his prodigious literary output for decades, both Tanner and James faded from prominence as Indian removal was accelerated.

Tanner and James were unusual cross-cultural collaborative partners. James was motivated in part by his advocacy for Native American rights, and in part by his zeal for scientific inquiry. Alone, this combination allowed James to present some of the most progressive arguments regarding the settlement and land use patterns he observed first-hand along the frontier. When he collaborated with Tanner, who was already nationally known for his unusual life story, the result was a captivity narrative unlike any other. The explicit challenge they issued may have fallen on mostly deaf ears at the time, but it never entirely faded away. Today, their work stands as a testimony to the cross-cultural resistance mounted against Indian removal.

"Rare or very local:" James's final decades

While James was only in his mid-thirties when he left Sault Ste. Marie, he never achieved great public success after the publication of Tanner's narrative. Several years were spent in Albany, New York, where he became involved in the temperance movement, especially after resigning his Army post in 1833. James had lobbied successfully to be transferred to Albany, New York, where he began working on temperance efforts immediately. Under mysterious circumstances, he was asked to resign his Army post in 1833 (Benson, "Edwin James" 280–85). By 1837, James had returned to the frontier, settling permanently on 320 acres outside of Burlington, Iowa. James had a brief and frustrating experience as an Indian agent at Council Bluffs, but maintained his farm until his death in 1861 (Benson, "Edwin James" 285–329).

James, while no longer traveling himself, remained painfully aware of the injustice caused by Native American relocation. Letters from the final decade of his life are markedly different from the high-spirited texts he produced in his twenties and thirties. While James's early work was influential among elite circles in the East—circles in which he had once moved quite ably—his retrospective self-assessments grew increasingly humble and melancholy after many years spent as a farmer and surveyor.[25] His efforts on behalf of social justice, however, never ceased. A highlight for James in later years was his farm's status as a stop on the Underground Railroad. The only fugitive slave case in Iowa's history dealt with a man arrested in James's presence. After a brief hearing, the accused fugitive, named Dick, was released: "more than a thousand exulting people escorted Dick to the ferry-boat upon which Dr. James, Dick and plenty of guards crossed the river [...] towards Chicago without detention" (Frazee 133).[26] James thus visibly demonstrated his commitment to issues of social justice throughout his life. In a letter he wrote to his niece in 1859, James wrote that "one race [is] as good as the other" (qtd. in Pammel 289).

Unfortunately, some commentators have evaluated James's lack of literary output in his final decades solely as a sign that he was embittered. This judgment may or may not be faulty, but it certainly fails to appreciate James's remarkable life. He never compromised his moral judgments and never stopped contributing to the struggle for justice on the American frontier. An especially touching and philosophical reply to why James's later output was slight is offered by his own 1854 answer to his botanist friend Torrey's questions about what James had been working on. James replied,

> As this world counts doing: *little* or *nothing*. It did not take me long to discover that it was not for me to make my mark upon the age and having settled that point to my own satisfaction I determined to make it on myself. I said "I will rule my own spirit and thus be greater than he that taketh the city" [...] to a true lover of nature like yourself, I will say no more about these things [...] It enters into my day dreams that I may yet go forth to gather weeds and stones and rubbish for the use of some who may value such things, and perhaps drop this life-wearied body beside some solitary stream in the wilderness.
>
> (qtd. in Pammel 178–79)

In the end, a comparison between Edwin James and an early description of the alpine shrub *Jamesia americana* rings true. Asa Gray and Torrey—whose work offers another fine example of successful collaboration—wrote an ambitious 1838 book describing all known flora in North America. Their description of *Jamesia americana* mentions that the plant is "probably rare or very local, as no other botanist seems to have met with it" (qtd. in Benson, "Edwin James" 126).

Because James's name has been removed from the landscape and even from his texts in some instances, a reappraisal of his contributions is necessary. Obviously, a contemporary renaming of Pikes Peak after James, the first man to record an ascent (and the man who was the mountain's namesake for many years) is not feasible. However, the mountain eventually named after James is an appropriate mountain to bear Edwin James's name. James Peak, near Idaho Springs, Colorado, stands 13,294 feet high and is the centerpiece of the James Peak Wilderness Area. A total of about 14,000 acres were designated as a federal wilderness area by the "James Peak Wilderness and Protection Area Act" (H.R. 1576) on August 21, 2002.[27] While the thousands of annual visitors to James Peak Wilderness do give passing attention to the legacy of James, renewing James's name and importance in literary and historical fields is equally important.

As for John Tanner, his name was given to Tanner Lake and Tanner Rapids, both of which are part of the Maligne River (called the Bad River in Tanner's narrative). Tanner reported:

> I was with great effort pushing up my canoe against the current [Tanner Rapids] which compelled me to keep very near the shore, when the

> discharge of a gun at my side arrested my progress. I heard a bullet whistle past my head, and felt my side touched, at the same instant that the paddle fell from my right hand, and the hand itself dropped powerless to my side. (270)

Today, the site of Tanner's attempted murder is located within Quetico Provincial Park, a wilderness area in Ontario that is more than one million acres in size. On the American side, Boundary Waters Canoe Area Wilderness is approximately equal in size.[28] Fittingly, Tanner Lake and Tanner Rapids are now counted as part of the only seven lakes and the Maligne River in Quetico on which Ojibwe members of the Lac La Croix Guides Association are allowed to run up to 10 h.p. motors (Keller 64–65). Part of a long list of primary and historical texts of the region, Tanner's narrative (or *The Falcon*) continues to be cited in places like canoeing.com's list of "Books and Media for the Advanced Paddler" and the National Park Service's list for Voyageur's National Park on the topic of the environment and the fur trade.[29] While the places named after Tanner are small in the larger scale of the roughly two million acres of protected land, Tanner's narrative continues to serve as inspiration for people interested in learning more about these landscapes.

Edwin James may be best known for his ascent of Pikes Peak and his official title as botanist for the Long Expedition. Yet James consistently advanced a forceful argument for environmental and social justice on the rapidly changing frontier of the early nineteenth century. Although he has been relegated to footnotes due in large part to his willingness to collaborate, I believe that his cross-cultural collaborations offer a valuable bridge between differences that many Americans once viewed as insurmountable. James's literary work and fascinating life story should be celebrated as a model of foresight which illuminates the connections between environment and ethnicity.

Notes

1. Benson's unpublished 1968 dissertation, "Edwin James: Scientist, Linguist, Humanitarian," remains the authoritative source of information on Edwin James, and includes a bibliography of all extant archival sources and letters known at the time of its writing.
2. The book is: John Tanner's *The Falcon: A Narrative of the Captivity and Adventures of John Tanner.* Ed. Louise Erdrich. New York: Penguin Books, 1994. Several editions of the book are discussed later in this chapter.
3. A source that may have influenced Williams is Susan Delano McKelvey, *Botanical Explorations of the Trans-Mississippi West, 1790–1850* (Jamaica Plain, MA: Arnold Arboretum of Harvard University, 1955).
4. John Torrey had a hand in the appointment by writing a recommendation for James (Benson, "Edwin James" 43–44).
5. George Miles's essay, "The Edwin James Letter Book at Yale," examines recently available letters to which Benson did not have access. These early letters, which

were purchased by the Yale Collection of Western Americana from a philatelist in 1983, begin in 1819 (Miles 35).

6. James's journal covers most of the 1820s, albeit with many gaps in time. Rotella's concerns about James's self-doubt and cynicism may be traced in some sections. Ironically, James recorded climbing Pike's Hill, in Minnesota, while studying geology.
7. See, for example, Robert Thacker's "The Plains Landscape and Descriptive Technique" (*Great Plains Quarterly* 2.3 [Summer 1982]: 146–56) and Herbert V. Fackler's "Cooper's Pawnees" (*American Notes & Queries* 6.2 [October 1967]: 21–22).
8. Bears are discussed on pages 52–57.
9. While Seymour's etchings were published as part of the narrative, Peale's drawings were not part of the final product. The American Philosophical Society in Philadelphia maintains a large Titian Ramsay Peale collection, including many works he created while part of Long's Expedition.
10. Discussion of bison appears on nearly every page from 461 to 488.
11. This comment appears in a letter from James to his brother, dated January 26, 1832. The publication is: James, Edwin and John Tanner. *Kekitchemanitomenahn gahbemahjeinnunk Jesus Christ, otoashke wawweendummahgawin.* Albany, New York: Packard & Van Benthuysen, 1833.
12. Most of the commentary has repeated the 1830 narrative's claims that the abduction of John Tanner occurred in 1789, and by Shawnees. However, John Fierst uses a deposition from 1790 to convincingly show that date to be more likely ("A 'Succession of Little Occurrences'" 16). In this 2012 article, Fierst also differentiates carefully between the Ojibwe group that took Tanner captive and the other Anishinaabe groups in the text (4–6). With help from John Nichols, who created retranscriptions of various individual's names as well as place names, Fierst has carefully reworked Edwin James's system (22, note 2; 23, note 10). However, the rest of my chapter uses the spellings that James used and have elsewhere been applied to the narrative.
13. Not available in the recent Penguin editions, the introduction by James is available in: John Tanner and Edwin James, M. D. *A Narrative of the Captivity and Adventures of John Tanner (U.S. Interpreter at the Saut de Ste. Marie) During Thirty Years Residence Among the Indians in the Interior of North America.* Ed. Noel Loomis. Minneapolis: Ross & Haines, 1956. Loomis's introduction accurately covers Tanner's later years, and this edition also reprints Part II of the original text. All quotations from the introduction and from Part II of the narrative come from this Ross & Haines edition; the quotations from Part I (Tanner's life story) are also from this edition (though the recent and more readily available Penguin editions are a facsimile version and so retain the same pagination).
14. Interestingly, the extensive passages and drawings describing "love medicine" in this chapter are an apparent source for Louise Erdrich's fiction, as detailed by Peter Beidler. Beidler also notes that Erdrich is "more willing than Tanner to accept as fact the magical ways of the Ojibwe" (38).
15. Fierst unequivocally states that "swallow" is the correct translation ("A 'Sucession of Little Occurrences'" 5).
16. For full details on the publication history, with the various titles under which the narrative has been published, see Appendix 2.
17. Namias, who explains the many permutations the Jemison text has undergone, eventually chose to republish Seaver's original document due to the impossibility

of identifying Jemison's own voice ("Editor's Introduction" 3–45). Margaret Fuller, in her reviews of various Native American studies, critiqued Schoolcraft by saying "a worse use could hardly have been made of such fine material" (20).

18. The 1830 review in *The Albion* does take some pains to emphasize the authenticity of the narrative, and contrast it with a captivity narrative published a few years earlier by a man named John Dunn Hunter, whom *The Albion* calls a "miserable swindler" and "a rogue" (163). Interestingly, Hunter's account has been accepted by recent literary scholars such as Namias as authentic, despite charges of fraud he faced during the years immediately following publication of his captivity narrative.
19. The narrative in question is: Henry Rowe Schoolcraft's *Narrative Journals of Travels from Detroit Northwest through the Great Chain of American Lakes to the Sources of the Mississippi River in the Year 1820* (Albany, New York: E. & E. Hosford, 1821).
20. Parker quotes a fascinating 1829 exchange between Henry and Jane on their ideas of gender. Henry wrote to his wife that "[i]t is the order of providence that man should be active, & women quiescent" (37). Jane replied, rather ironically, by citing the newspapers in which she can follow him more readily than in his occasional letters to her, "I must learn that great virtue of a woman—quiescence" (38). Mielke notes that she was unable to fully integrate all of Parker's scholarship due to her own publishing deadlines, but stands by her argument regarding familial collaboration (230).
21. Its full title is *The Holy Bible: Containing the Old Testament and the New. Translated into the Indian Language.* (Cambridge, Massachusetts: Samuel Green and Marmaduke Johnson, 1663).
22. Lepore's *The Name of War* discusses James Printer at length in the chapter entitled "Come Go Along with Us" (125–49). Her thesis for this chapter is that "[t]he lasting legacy of Mary Rowlandson's dramatic, eloquent, and fantastically popular narrative of captivity and redemption is the nearly complete veil it has unwittingly placed over the experiences of bondage endured by Algonquian Indians during King Philip's War" (126). In his 1997 edition of Rowlandson's narrative, Salisbury reprints a letter negotiating the terms of Mary Rowlandson's release and ransom, which James Printer wrote as scribe.
23. Fierst's article "Return to Civilization" does an excellent job with this phase of Tanner's life, turning up previously unknown documents and offering compelling evidence that Tanner had been framed for the murder of James Schoolcraft.
24. One account that seems less than trustworthy, but is consistent with Schoolcraft's assessment of Tanner, is Dr. Chas. A. Lee's "The Residence of Tanner, Or the Indian Whiteman" (*Dwights American Magazine, and Family Newspaper* 2.26 [July 25, 1846]: 388). Lee writes in response to the murder of James Schoolcraft, and attempts to show that Tanner was not only capable of the crime, but had perhaps been contemplating it.
25. Many fascinating letters written by James during his Burlington, Iowa, years were borrowed from family members and reprinted in their entirety in 1907 and 1908 by L. H. Pammel. Pammel's first installment is also the source for the only known extant image of James, which was "[f]rom a miniature on ivory in possession of the family" (Pammel 160).
26. Details about James and the fugitive slave, Dick, are discussed by George Frazee, Commissioner of the US Court in Burlington, who wrote a detailed article about the event. See Frazee's "The Iowa Fugitive Slave Case" (*Annals of Iowa* 4.2

[July 1899]: 118–37). Frazee also suggests that one reason some viewed James as eccentric was due to his well-known stance in favor of abolition. See also William Salter's *The Life of James W. Grimes, Governor of Iowa, 1854–1858; A Senator of the United States, 1859–1869* (New York: D. Appleton and Co., 1876). Page 73 describes the Iowa fugitive slave case and compliments James for his role in it.

27. The official federal website for the James Peak Wilderness is: www.fs.fed.us/r2/arnf/recreation/wilderness/jamespeak/index.shtml.
28. The official websites for Quetico and the BWCAW are: http://www.ontarioparks.com/ENGLISH/quet.html and http://www.fs.fed.us/r9/forests/superior/bwcaw/.
29. The websites are: http://www.canoeing.com/advanced/booksmedia/ and http://www.nps.gov/parkhistory/online_books/voya/futr/biblio.htm.

4 All along the Watch Tower

Life of Black Hawk as a counter-captivity narrative

The term "counter-captivity narrative" has been used by noted captivity scholar Kathryn Zabelle Derounian-Stodola to describe a text that "thwarts, appropriates, and modifies elements of the conventionally Western genre" of the Indian captivity narrative (161). The counter-captivity narrative genre is exemplified by texts produced by Native American authors who use the conventions of the Indian captivity narrative in the service of challenging cultural stereotypes of Native Americans. Her major examples are primarily by contemporary writers, including Virginia Driving Hawk Sneve, Sherman Alexie, Louise Erdrich, and Gerald Vizenor, but I argue in this chapter that *Life of Black Hawk* is an example of an early counter-captivity narrative. I examine the scholarship, textual history, and language of the narrative to argue that the text has at least as many commonalities with the tradition and genre of the captivity narrative as it does with the autobiography.

In addition to *Life of Black Hawk*, Native American-authored texts as widely divergent as *Geronimo's Story of His Life* (1906) and William Apess's *A Son of the Forest* (1831) may be seen as early counter-captivity narratives. Apess, interestingly, turns captivity on its head. Apess wryly admits that as a child, his greatest fear was that he would be "sent away among the Indians into the dreary woods" (10); this fear was due to the "stories I had heard of their cruelty towards the whites—how they were in the habit of killing and scalping men, women, and children" (11). As Gordon Sayre notes, both Black Hawk and Geronimo underwent a "process of war, defeat, and captivity" that exposed the men to white American societal expectations (*Indian Chief* 8). Sayre notably closes his 2000 anthology *American Captivity Narratives* with an excerpt from *Geronimo's Story of His Life*. While slave narratives are widely understood to be written by black authors, Indian captivity narratives are in these cases written by Native Americans and can be called counter-captivity narratives. Just as slave narratives were often used to advance abolitionist goals, counter-captivity narratives were used to resist personal and tribal relocation.

Historians and critics from 1834 to the present have generally considered *The Life of Black Hawk* as an autobiography, focusing on the idea of Black Hawk as a tragic hero and as the author-creator of an entirely new genre of ethnic literature. In this view, the compositional collaboration of Black Hawk, LeClaire, and Patterson is castigated for compromising the autobiographical authenticity of *Life of Black Hawk*; however, critical attempts to "isolate" Black Hawk's contributions to the text only reinforce stereotypes of the pure Indian who is contaminated and destroyed through contact with whites. Instead, as Kathleen Boardman observes about collaborative life narratives, cross-cultural collaboration is "an important practice, a potential force for diversity and even for justice. ... [I]t is a means for voices of resistance to mainstream values to be heard" (177). I argue that we can gain a more accurate and more positive understanding of *Life of Black Hawk* through an acceptance of its collaborative composition. The publication of a statement of active resistance to Indian relocation was only possible, after all, through the combined efforts of Black Hawk, LeClaire, and Patterson.

In 1832, Black Hawk led at least 1500 Native Americans across the Mississippi from west to east, leaving what is now Iowa for Illinois, in a bid to reinhabit traditional Sauk Indian lands along the Rock River. This act of resistance to white encroachment was one of many in a long dispute over an 1804 treaty that allegedly ceded all lands in Wisconsin and Illinois to the US government. Black Hawk, believed to have been sixty-six years old at the time of the crossing of the Mississippi, was known to some as a military tactician seasoned by his involvement in the War of 1812, but he was far from nationally prominent. Yet Black Hawk's name became a fixture in national news for months and even years to come, as the Black Hawk War began under dubious circumstances in May of 1832—when a white militia killed several Sauks who were carrying a white flag of truce—and continued until early August of that year.

Upon his band's defeat in the massacre known as the Battle of Bad Axe, Black Hawk was captured and brought eastward in chains to meet with President Andrew Jackson, after which he and several other prisoners were compelled to go on a tour of major cities of the eastern US. This odd decision on the part of the US government was reportedly made in order to demonstrate to Black Hawk the futility of armed resistance. Yet, of course, Black Hawk knew this firsthand, as he was already held by the government. Perhaps more likely, the government hoped to persuade whites that the Black Hawk War had been the "Indian war to end all Indian wars" and that Black Hawk's tour of the East would signal a long-term cessation of frontier disputes. This supposedly rational argument was applied by William Snelling, for one, in his 1835 review of *Life of Black Hawk*. Snelling wrote that

> Indians ... cannot but see that resistance against any claim of the United States, just or unjust, is the very worst policy they can pursue. Hereafter, it is probable, that whenever a wish is expressed to remove them from their lands, they will readily comply on the best terms they can make, since any resistance they can offer will but protract the catastrophe and render it more appalling (85).

Counter to this expectation, the tour failed to convince Black Hawk that his career was over. Black Hawk and the other Sauks with him drew immense crowds, most of whom were quite sympathetic toward them. When Black Hawk was returned a year later to the West, he was expected to be subordinate to Keokuck, a Sauk leader Black Hawk felt had been an apologist and a collaborator with the US government. Black Hawk decided to publish a book to tell his side—and his people's side—of the story. In August 1833, Black Hawk dictated his life story to a mixed-raced translator named Antoine LeClaire, who worked at the Rock Island Indian Agency, and the transcript was then reworked by newspaper editor John B. Patterson. The resulting text was finished by mid-October 1833, according to Le Claire's note of certification (5).[1] Initially published in Cincinnati in 1833 under the title *Life of Black Hawk*, the book was expanded and reissued by Patterson with the title *Autobiography of Black Hawk* in 1882. *Life of Black Hawk* was both popular and influential, with reviews of it published in important journals such as the *North American Review*, and it was reprinted in several cities in 1834.

More than half of *Life of Black Hawk* takes place before the Black Hawk War. These pages include much ethnographic information, including Sauk cultural traditions, as well as detailed recollections of Black Hawk's personal history. These personal details center on various battles and allegiances Black Hawk engaged with. Black Hawk also takes pains to outline his claim to ancestral lands, at times using a discourse mode similar to that of treaties. The remainder describes the Black Hawk War from Black Hawk's perspective and closes with Black Hawk's captivity and experiences in the East.

An examination of contemporary visual depictions of Black Hawk when he was a prisoner at Jefferson Barracks immediately following the conclusion of the war serves as a valuable summary of my overall argument. These images show that he saw himself as a resisting captive, engaged with the tradition of captivity, if not necessarily with all the literary tropes of the genre. The famous American portrait and landscape painter George Catlin traveled to Jefferson Barracks specifically to paint portraits of Black Hawk and other Native American captives. Catlin is often celebrated as having been sympathetic, if not progressive, in his treatment of Native Americans on the frontier. In both images, Black Hawk is readily identified by the fan he holds, created from the feathers of the bird that is his namesake.[2] Catlin's 1832 portrait of Black Hawk, entitled "Múk-a-tah-mish-o-káh-kaik, Black Hawk, Prominent Sac Chief," helped create a romanticized legacy of Black Hawk as an autobiographer and tragic hero—a legacy I wish to complicate. In this portrait, reproduced here as Figure 4.1, Black Hawk is depicted as a solitary chief, divorced from both his homeland and from his situation as a prisoner. This melancholic, dignified image of Black Hawk from the waist up seems to ask for our sympathy, but the events of the Black Hawk War—at the time only weeks old—are already distant and irreconcilable because of Black Hawk's apparent surrender to American encroachment.

Manifest Destiny is thus reinforced rather than questioned, despite the victimization of Black Hawk, who apparently will fade away, dignity intact, all else lost. Kerry Trask describes this and other portraits of Black Hawk as helping to make Black Hawk "a reflection of some of American culture's most powerful myths and intriguing fantasies, evoking strong feelings of both admiration and pathos, which gave white America a bittersweet sense of heroic tragedy" (298). Interestingly, Timothy Sweet argues that Catlin's 1832 portrait feminizes Black Hawk, "as if he were an elegant lady holding a fan" (489). Sweet's implication is that Black Hawk's rival Keokuck is therefore anointed as the new warrior-chief of the Sauks, as he was painted in a traditional masculine pose, with his horse and tomahawk.

In an additional image that was published in 1833 by *The New York Mirror*, Black Hawk is depicted in a knee-length, buttoned-up coat, holding a peace pipe as well as the hawk-fan. This coat can be considered accurately as his prison uniform. *The Mirror* noted Black Hawk's refusal to hold a spear for Catlin, relating that Black Hawk was "indignant" and said

Figure 4.1 "Múk-a-tah-mish-o-káh-kaik, Black Hawk, Prominent Sac Chief." George Catlin, 1832. Smithsonian American Art Museum. Gift of Mrs. Joseph Harrison, Jr.

"[n]o spear for me! I have forever done with spears!" (Anon. 9). Despite their status as prisoners, the agency of Black Hawk and the other Sauks should not be understated. Under pressure from the captive Sauks he depicted, Catlin also painted a lesser-known group portrait that showed the men, including Black Hawk, in chains (Figure 4.2).

Figure 4.2 "Black Hawk and Five Other Saukie Prisoners." George Catlin, 1861/1869. Paul Mellon Collection. Courtesy of the Board of Trustees, National Gallery of Art, Washington, DC.

Neapope refused to hold the peace pipe while Catlin painted this image. The *Mirror* reported that he instead picked up his ball and chain saying, "Paint that! and let the Americans see that they have Nah-ope—a prisoner, and in irons!" (9). In 1838, Benjamin Drake noted that Neapope also "kept varying his countenances with grimaces, to prevent [Catlin] from catching a likeness" (202–203). Clearly, the captives were not content with posing for Catlin as if they were free warriors. Rather, they desired to be depicted accurately, as prisoners of the US Army. The painting "Black Hawk and Five Other Saukie Prisoners" suggests that Catlin has widened his frame, pulling back to reveal the full context of what he saw at Jefferson Barracks. The six men are shown looking in various directions, each holding his ball and chains aloft, as if searching for a more reliable witness to their condition. John Hausdoerffer notes that this is not a stylized depiction of the noble savage as in the first painting, but rather "a near-documentary representation of their prison conditions" (116). Nonetheless, Neapope's reported grimaces are not depicted here by Catlin, who instead shows the men with the same placid countenances that typify his individual portraits; perhaps he was attempting to mitigate against the continued resistance and defiance of the Sauks. Catlin did not finish this painting for several decades, while the first painting was displayed widely and has been an iconic image since 1832. The shackles are also emphasized in *Life of Black Hawk*:

> I felt the humiliation of my situation: a little while before, I had been leader of my braves, now I was a prisoner of war! [...] We were now confined to the barracks and forced to wear the ball and chain! This was extremely mortifying, and altogether useless. [...] Having been accustomed, throughout a long life, to roam the forests o'er—to go and come at liberty—confinement, and under such circumstances, could not be less than torture! (79)

Due to the agency of Black Hawk and others, it is clear that the later Catlin painting provides the more accurate depiction, one reflecting the desire on the part of the Sauks to be portrayed as resisting captives rather than as noble chiefs, fading away.

I believe *Life of Black Hawk* should not be generically categorized as an autobiography. My larger argument is paralleled by the differences between the two Catlin portraits. The view that Black Hawk's narrative is an autobiography has been extremely persistent—as I detail later, the editors of the five major editions published in the past one hundred years all consider it an autobiography—but I believe this judgment on the genre of *Life of Black Hawk* is akin to looking at the Catlin portrait as if Black Hawk were indeed a romantic hero. Rather, we should also gaze upon the later painting, which argues that *Life of Black Hawk* is a captivity narrative, a contribution to a genre that was very popular in the early nineteenth century and was then already 150 years old. Indeed, no text that is so deeply a cross-cultural collaboration as *Life of Black Hawk* can truly be understood primarily as an autobiography, a genre conditioned upon ostensible single authorship. While the text may be seen as an early example of a collaborative life narrative, a genre about which scholars are now rapidly gaining new awareness, also considering it as a captivity narrative allows us to engage with the text on historical and literary terms that were influential during the nineteenth century. Captivity narratives were routinely produced to arouse popular sentiment regarding cultural contact zones, and the contested nature of the frontier was played out for two hundred years in the pages of these texts. *Life of Black Hawk* shows a concerted effort by Black Hawk to clear his name and exercise influence regarding his fate and that of his people immediately upon his release from captivity.

While Black Hawk's text is not normally considered a captivity narrative, its original title includes the phrase "his surrender and confinement at Jefferson barracks." While the captivity occurs late in *Life of Black Hawk*, by being taken captive and rushed across the country on a lengthy tour,

Black Hawk's story mirrors that of many captives who emphasize the disorienting journeys that begin a period of captivity. Some commonalities that *Life of Black Hawk* shares with many captivity narratives include the nature of Black Hawk's wartime imprisonment; the acculturation he develops with his captors; the related decision to publicly state his defense; and, perhaps most importantly, the complex cross-cultural collaboration necessary for publication. So my argument is not only that *Life of Black Hawk* has much in common with captivity narratives, but that the captivity narrative was a malleable genre that Black Hawk used in deliberate and strategic ways. By reading *Life of Black Hawk* as a cross-cultural captivity narrative, we can better understand its authorial intentions and its related message of environmental justice.

The environmental message this chapter closes with is also related to the Catlin portraits. Just as Black Hawk and the other captives wished their captive condition to be depicted accurately, so too does Black Hawk wish to emphasize setting within his narrative. His most detailed descriptions do not appear during battles, nor are characters other than Black Hawk himself especially well-developed in the narrative; rather, the text is replete with descriptions of the fertile mosaic of fields, forests, and waterways that the Sauks called home. If Catlin seems to widen his perspective to capture the full scene in the later painting, Black Hawk similarly widens the scope of his own narrative. My argument here is twofold. First, Black Hawk details the land carefully in order to show his deep love for a specific place—a rationale given both by Black Hawk and by many later commentators for his willingness to fight to defend Saukenuk, a homeland centered where the Rock River enters the Mississippi, at what is now called the Quad Cities.[3] Black Hawk's description of Saukenuk is closer to a pastoral landscape than a wilderness, as he emphasizes that this landscape was both hospitable and actively inhabited:

> We had about eight hundred acres in cultivation [...]. The land around our village, uncultivated, was covered with blue-grass, which made excellent pasture for our horses. Several fine springs broke out of the bluff, near by, from which we were supplied with good water. The rapids of Rock River furnished us with an abundance of excellent fish, and the land, being good, never failed to produce good crops of corn, beans, pumpkins, and squashes. [...] Our village was healthy, and there was no place in the country possessing such advantages. (42)

With about six thousand Sauk residents during the summer months, Saukenuk was larger than Chicago during the first third of the nineteenth century. The environmental history of Saukenuk suggests that the publication of *Life of Black Hawk* has had lasting ramifications for the land today called Black Hawk Memorial Site. But before discussing that particular place, a closer look at *Life of Black Hawk* will help to demonstrate why the text has been so influential.

Black Hawk's narrative went through five printings in 1833 and 1834 alone. His depiction of his own captivity during the concluding section of the narrative served as a familiar trope to readers and also previewed his close attention to the landscape:

> On our way down [the Mississippi River], I surveyed the country that had cost us so much trouble, anxiety, and blood, and that now caused me to be a prisoner of war. [... I] recollected that all this land had been ours, for which me and my people had never received a dollar, and that the whites were not satisfied until they took our village and our graveyards from us, and removed us across the Mississippi. (79)

Such passages would have been familiar to readers of captivity narratives, who were accustomed to accounts of the depredations of the often fictionalized "savages" that killed and ravaged whites. Yet the eighteenth-century conventions of captivity narratives had shifted along with the westward push of white settlement, leaving a disconnect that readers in the eastern US felt in their hunger to read more diverse accounts of contact between whites and Native Americans. The inversion of agency in Black Hawk's narrative is striking. Rather than a tale of a white woman being made captive of Native Americans, *Life of Black Hawk* is the story of a Native American man resisting white settlers and made a captive of the US Army. This text was read sympathetically because readers in the East enjoyed a high degree of security in the early nineteenth century. Reading Black Hawk's narrative allowed them the vicarious thrill of imagining themselves in his position, a thrill that they were accustomed to experiencing through reading the accounts of more conventional captivities.

Despite such conventional captivity narratives, which often featured a "damsel in distress," by 1830, cultural space had been cleared for alternative storylines, as seen in the captivity narratives of John Dunn Hunter (1823), Mary Jemison (1824), and John Tanner (1830), to name three so-called "white Indians" who were kidnapped at a young age and became fully acculturated to Native society. Although none was conventionally "redeemed" through a return to mainstream white society, their narratives were widely read by national and even international audiences in the decade immediately prior to the publication of *Life of Black Hawk*. Fiction of this decade also chipped away at the racial boundaries, ranging from Cooper's *Leatherstocking* novels to Sedgwick's *Hope Leslie*. Thus, imaginative readers must have seen the connection of skin color and culture as potentially alterable. Transculturated white Indians were rarely depicted in positive terms before this time, but a "different impulse began to dominate the publication of narratives in the East" by the 1820s and 1830s (Derounian-Stodola, Levernier 35). Captivity narratives no longer demonized Native Americans, but instead became mainstream literary opportunities to record national history and scientific information as well as ethnographic details and to examine "nationalistic assumptions about progress, race, and Manifest Destiny" (Derounian-Stodola, Levernier 36–37). By sympathizing with Black Hawk during his tour and through the reading of *Life of Black Hawk*, sizable Eastern audiences followed this alternative, and quickly growing, tradition of counter-captivity narratives.

While Black Hawk's own experience as a prisoner can be seen as an example of a counter-captivity, many additional captivities are detailed in *Life of Black Hawk*. In the opening pages, Black Hawk engages in a traditional storytelling mode, the counting of coups: "We [...] killed all [...] inhabitants [of a large Osage encampment], except two *squaws!* whom I captured and made prisoners" (15; emphasis in original). Black Hawk, in a separate event against Cherokees, explains that he wished to engage a large group in battle, but finding only five individuals, whom he subsequently imprisoned, states that "I afterwards released four men—the other, a young *squaw*, we brought home" (15; emphasis in original). While white readers were very familiar with the Indian captivity narrative, Native Americans were equally familiar with captivity practices and stories, having traditionally taken captives—both from other tribes as well as white captives—to replace dead relatives and for trade.

White captivity is also integrated into Black Hawk's narrative, most notably in the story of teenagers Sylvia and Rachel Hall, who were taken captive by Potawatomi Indians affiliated with Black Hawk's band during the Black Hawk War and later released by Black Hawk. Black Hawk takes pains to tell his white audience that the Hall sisters were not mistreated by the Sauk:

> The Pottowatomies killed the whole family, except two young squaws, whom the Sacs took up on their horses, and carried off to save their lives. –They were brought to our encampment, and a messenger sent to the Winnebagoes, as they were *friendly on both sides*, to come and get them, and carry them to the whites (74; dash and emphasis in original).

The hand of Patterson as editor may be seen here, as the term "squaw" is applied to virtually every female character in the text, whether young or old, white or Native American. Despite the shortcomings of Patterson, his editorial decisions are not egregious by the standards of Western US regional writers. For example, a brief book based upon the captivity of the Hall sisters, published in order to further inflame local sentiment against Native Americans, shows that some American audiences remained susceptible to stereotypical depictions of Native Americans:

> Since the commencement of hostilities by the Sacs and Foxes, and in the many depredations committed upon the defenceless inhabitants of the frontier settlement, the lives of but few, who have been so unfortunate as to fall into their hands, have been spared. Their tomahawks have, literally, been made drunk with innocent blood! the virgin's

> shriek, the mother's wail, and the infant's trembling cry, has proved music in their ears!
>
> (Edwards 17)

This melodramatic account attempts to justify the actions taken by white militias, and unconvincingly obscures the fact that the Hall sisters were returned, unharmed, after approximately ten days in captivity. Decades later the sisters testified that they were treated humanely and were well fed by the Sauks: "The Indians then prepared their supper, consisting of dried meat sliced, coffee boiled in a copper kettle, corn pounded and made in a kind of soup; they then gave us some of this preparation in wooden bowls, with wooden ladles" (qtd. in Stevens 151).[4]

Food is often a key part of captivity narratives, and interestingly the Halls depicted a well-rounded meal that was cooked and served in a "civilized" manner. Clearly the Sauks were adequately provisioned at the outset of the Black Hawk War, though they suffered a great deal as the summer progressed. Black Hawk relates that hunger motivated his attempted return to the rest of the Sauk nation and forced his engagement with the white militia at the Battle of Wisconsin Heights. His comments might have come straight from Mary Rowlandson's narrative: "We were forced to *dig roots* and *bark trees*, to obtain something to satisfy hunger and keep us alive! Several of our old people became so much reduced, as actually to *die with hunger!*" (74; emphasis in original).

Before turning to a description of the battle, however, Black Hawk takes pains to emphasize his good treatment of white captives, and this care is apparent even to two teenagers who are traumatized by witnessing the killings of most of their family. Decades later, they did not make any pejorative comments about Black Hawk himself. I believe Black Hawk was well aware of the conventions and expectations of Indian captivity, if not the genre of the captivity narrative itself, and was actively shaping his public image in his good treatment of the Hall sisters, as well as in the writing of *Life of Black Hawk*.

Life of Black Hawk supplanted by *Black Hawk's Autobiography*

The *North American Review* suggested in 1835 that the text first published as *Life of Black Hawk* was a "curiosity; an anomaly in literature. It is the only autobiography of an Indian extant, for we do not consider Mr. Apes [William Apess] and a few other persons of unmixed Indian blood, who have written books, to be Indians" (Snelling 68). This rather startling definition suggests why *Life of Black Hawk* is often considered the first Native American autobiography: to many scholars, the act of gaining literacy seemed to remove an essential element of "Indianness." Of course, publishing an autobiography is nearly impossible without the ability to write. Snelling continues, "here is an autobiography of a wild, unadulterated savage, gall yet fermenting in his veins, his heart still burning with the sense of wrong [...] and his hands still reeking with recent slaughter" (69). Snelling suggests that an accurate way of reading Black Hawk included emphasizing the idea of Indians as savages, thus allowing readers to imagine themselves in Black Hawk's situation without necessarily having to sympathize with his actions. In this interpretation, Black Hawk was romanticized as an exemplary "chief" whose decisions were primarily influenced by his instincts. This view suggests that race is a measurable category and cannot be overcome by environmental or social factors. Snelling was hardly alone in this assessment of Black Hawk; for example, Black Hawk was the subject of a lengthy phrenological study in 1838 that analyzed the shape of his skull based on measurements from a plaster bust. On the one hand, the study found his head to be "large, giving much more than an ordinary amount of intellect and feeling," yet on the other, certain "organs, when large ... would make an Indian the bold and desperate warrior" (Anon., "Phrenological Developments" 32, 33).

Arnold Krupat noted in his foundational work on American Indian autobiography that Black Hawk's narrative is the "unprecedented instance of an Indian speaking for himself" (*For Those* 50). Krupat does admit, however, that "Indian autobiography is a contradiction in terms. Indian autobiographies are collaborative efforts, jointly produced by some white who translates, transcribes, compiles, edits, interprets, polishes [...]. I may now state the principle constituting the Indian autobiography as genre as the principle of *original bicultural composite composition*" (*For Those* 30–31). Despite qualifying his claim, Krupat ultimately adheres to a reading of *Life of Black Hawk* primarily as an autobiography. Yet because Black Hawk became significant and well-known in large part due to his captive tour of the East, I argue that *Life of Black Hawk* is a product of that historical moment in such a foundational way that the popular genre of the captivity narrative is best able to contain the motivations of the cross-cultural collaboration.

Nonetheless, every editor of the text for the past century has suggested that the book is an autobiography. James D. Rishell, in 1912, claimed that there is "no other book in existence with this peculiar ethnic quality, nor can there be another. ... the last Indian unmodified by the white man's schools ... and other strong influences of civilization, has passed away" (x). Milo Milton Quaife wrote in 1916 that "Whether true or untrue in its statements, and in this respect it shares the errors common to all autobiography, the book is important because it illuminates ... *the viewpoint and state of mind of a typical representative of the vanquished race*" (14; emphasis in original). Donald Jackson's comprehensive and scholarly edition, reissued several times between 1955 and 1990, concludes that the text is "basically a tale told by an Indian from an Indian point of view" (30). Roger L. Nichols wrote in 1999 that "[d]espite the editor's meddling with the account ..., the resulting prose still gives obvious examples of Sauk cultural practices and the warrior's individual attitudes" (xix). Despite the collaboration necessary for its creation, which is treated as a factor that taints its authenticity, the appraisal of the text as kicking off a new genre of American Indian autobiography continues unabated with J. Gerald Kennedy's introduction to the new Penguin edition, which was published in 2008. Still, Kennedy's version is the first major edition since 1916 to drop the word "autobiography" from the title. Yet for the first fifty years of the text's existence—the years closest to the events described within *Life of Black Hawk*—the title and presentation of the book was most accurate. Collaboration and contact between whites and Native Americans was sadly seen as less possible and less valuable as the decades progressed, and so editors redoubled efforts to emphasize the authenticity of Black Hawk's words as if Patterson and LeClaire had not contributed to the project.

Interestingly, the 1882 edition that titled the text as an autobiography was the only expanded edition of Black Hawk's narrative. Patterson published this edition in his old age, making many changes and adding stories that are certainly his own and not Black Hawk's. Because of this, virtually all reprintings have been of the 1833 or 1834 editions, which are seen as more fully reflecting Black Hawk's intentions. For example, the 1882 edition features a clearly fabricated romantic story of a Sioux brave who fell in love with a Sauk maiden; when the pair took shelter during a thunderstorm at Black Hawk's Watch Tower (a rock formation at Rock Island), the cliff they were under was struck by lightning, resulting in the lovers' ethno-mythical entombment under the shattered rock (65–66). After the conclusion of the text he called an autobiography, Patterson also added ancillary materials including additional wartime documents and historical sketches of various Mississippi River towns that were founded in the aftermath of the Black Hawk War.

Most notably from a narrative and stylistic point of view, Patterson added an embellished captivity narrative he wrote about Elijah Kilbourn called "A Reminiscence of Black Hawk" (159–69). According to this interpolated narrative, Kilbourn was a captive for three years, beginning just before the War of 1812, as well as a short time during the Black Hawk War. This brief captivity narrative shows Kilbourn as a captive of Black Hawk himself, not once, but twice. In the first captivity, Kilbourn is adopted into the Sauk tribe but eventually escapes; in the second, Kilbourn is delivered from captivity by Black Hawk personally. Kilbourn's story includes a stylized monologue detailing Black Hawk's wrongs at the hands of the whites—in other words, a summary of the main body of *Life of Black Hawk*, though Kilbourn himself never appears in the body. Many of the details of Kilbourn's story are suspect, and Patterson begins the section by writing that Kilbourn "loves to recount" these stories, which suggests that this narrative was the product of two then-elderly men embellishing frontier stories that had happened fifty years prior (159). Nonetheless, Kilbourn was certainly a member of the white militia that attacked Black Hawk's band.

Importantly for my argument that *Life of Black Hawk* is a counter-captivity narrative, Patterson independently recognizes the suitability of the captivity narrative to relate Black Hawk's point of view, even as he collaborates further in this ancillary narrative with Elijah Kilbourn "in whose language the story shall be given" (159). As demonstrated in Derounian-Stodola's *The War in Words: Reading the Dakota Conflict Through the Captivity Literature*, the historical backdrop of a war between whites and Native Americans often gave rise to associated captivity narratives that derive much of their importance and significance from their context. The various layers of captivity narratives placed within the larger counter-captivity narrative *Life of Black Hawk* are an indication of the great cultural power contained within the genre.

With all of these editorial notes from various editions in mind, and with the clearly corrupted 1882 edition functioning as a cautionary tale showing how easily Patterson added materials to the original edition, much of the scholarship on Black Hawk over the past twenty-five years has focused on discerning what portion of the narrative was solely Black Hawk's. Indeed, Nichols devotes much of his 1999 introduction to explaining what most critics have argued; his summary of the scholarship is that there is a "near-consensus … that Black Hawk's attitudes and ideas come through clearly despite the editor's apparent efforts to modify or even ignore his actual words" (xx). One of the few dissenters is Mark Wallace, who notes the usage of "'quaint' Indian colloquialism" such as "bad medicine" and "squaws" and concludes that "[a]long with this racist translation, the translation of Black Hawk's religious and political values may be similarly biased" (491). Wallace's argument, which in some ways is similar to my own, is that *Life of Black Hawk* is not an autobiography. Wallace accurately writes that "*An Autobiography* presents a cross-culturally produced 'Indian' voice that is responding to white voices in the language of the whites" (485). Yet Wallace misses an opportunity to do more with the text: the primary conclusion he draws from this insight is that Black Hawk's text is corrupt and in the service of a racist agenda, and therefore is inauthentic. I argue that rather than rejecting *Life of Black Hawk* as flawed, the cross-cultural collaboration that was necessary for its production has the positive result of presenting a voice of resistance that otherwise would not have been heard. David Brumble suggests that Patterson is an example of the "Absent Editor" who attempts to remove himself from the autobiography (75); yet, Brumble explicitly wants to "allow some pretty good guesses about where the Indian leaves off and the Anglo begins" (12). Ultimately, scholarship on Black Hawk has remained in service of a theory of autobiography reliant on the genre expectation of single authorship. In this view, the more of editor Patterson and translator LeClaire that can be identified and deleted, the better. By isolating Black Hawk in this way—insisting on his pure, Indian voice—we are unnecessarily denigrating the potential of the three men to work together.

Another problem with undervaluing the ability of Black Hawk to engage in a sophisticated way with his lifelong experience with whites on the frontier is seen in assumptions that he is somehow abused as a tool of his collaborative partners and is uniformly taken advantage of. For example, many critics have found the final lines of *Life of Black Hawk* to ring hollow: Black Hawk writes, "The tomahawk is buried forever! [...] may the watchword between the Americans and Sacs and Foxes ever be—'*Friendship!*'" (88). Yet two critics have recently disagreed, finding Black Hawk's treatment of peace medals to be richly ironic. With regard to the final words of the narrative, I find Kendall Johnson's article persuasive. Johnson argues that Black Hawk is cleverly alluding to US peace medals that often bore the phrase "Peace and Friendship." Mark Rifkin offers a similar and more sustained argument, examining the ways Black Hawk uses medals throughout the text to signify his allegiances. Black Hawk accepts a British medal during the War of 1812, for example, explaining that Robert Dixon "placed a medal round my neck, and gave me a paper, ... and a silk flag" (Black Hawk 25). Later in the narrative, Black Hawk angrily refuses to trade his British medals for US medals (83). That some artists, such as Charles Bird King in 1837, depicted him wearing a Martin Van Buren medal is ironic but perhaps accurate, as President Jackson did give Black Hawk such medals during their 1832 meeting.

Determining an underlying voice of Black Hawk that can be divorced from what Patterson and LeClaire may have altered or added is impossible. I argue that attempts to make such distinctions are also unnecessary, although learning as much as possible about the collaborative arrangement remains important to a full understanding of the narrative. The best evidence we have about how the text was created comes from the much-maligned 1882 edition. Patterson immediately follows the conclusion of the narrative with this note:

> After we had finished his autobiography the interpreter read it over to him [Black Hawk] carefully, and explained it thoroughly, so that he might make any needed corrections, by adding to, or taking from the narrations; but he did not desire to change it in any material manner. He said, "It contained nothing but the truth, and that it was his desire that the white people in the big villages he had visited should know how badly he had been treated, and the reason that had impelled him to act as he had done." (129)

A number of conclusions may be drawn from this passage. The first is that the methodology of composition—assuming it is accurately described—is both quite sound ethically and clearly collaborative. Black Hawk is involved even in the revision process. Audience concerns are clearly stated as well. Black Hawk is sharing his story with the whites in cities such as Philadelphia, Washington, and New York, and the book is thus not primarily intended for a local, white frontier audience. Finally, I find it telling that Patterson does not write that Black Hawk finished his autobiography; rather, he writes that "*we* had finished the autobiography" (129; emphasis added).

Theodore Rios's and Kathleen Sands's book *Telling a Good One: The Process of a Native American Collaborative Biography* helps show why the idea of "as told to" autobiographies is so problematic. Sands writes about her collaborative biography with Rios, a Tohono O'odham man she first interviewed in 1974. Over time, she realizes that she had been naïve to think she could write an autobiography of his life. Sands reflects that

> [p]recisely because of the collaborative nature of Native American orally narrated life story, forcing it into a Euro-American literary genre wrests the narrative out of the cross-cultural process from which it has been generated. … What's left of the term auto-bio-graphy, then is 'bio' (3).

Despite the cultural and period differences, Sands's argument about Rios also applies to the relationship between Black Hawk, LeClaire, and Patterson. She eventually decides on the term "collaborative biography" (77) because Rios's story is produced through dual authorship and is cross-cultural; Rios also "maintains his culturally determined methods of self-presentation" (70).

Self-presentation is a crucial part of Black Hawk's story as well. In addition to his successful efforts to influence how Catlin depicted him, Black Hawk also decided which peace medals were appropriate for him to receive or display and which were not. Black Hawk repeatedly explains the great importance of his medicine bag, valuing it as a repository of his people's traditions. Neil Schmitz closes his chapter "Black Hawk and Indian Irony" by noting that upon his return from the East, Black Hawk immediately petitioned for his medicine bag: "now, that I was to enjoy my liberty again, I was anxious to get it, that I might hand it down to my nation unsullied" (Black Hawk 85). His medicine bag, of course, was an important part of his traditional cultural identity. With this in mind, Schmitz immediately pivots to his final sentence of the chapter, echoing Black Hawk's diction by claiming that "it is for us to judge how altered it [Black Hawk's Sauk history] is, how sullied, by the devices of Patterson and LeClaire" (85). Yet very little is known of these "devices." Furthermore, suggesting that collaboration leads inexorably to contamination disregards the choices Black Hawk made throughout his life. Fixing a romanticized identity on Black Hawk ignores his long and varied life, and suggests that cross-cultural exchanges are uniformly damaging to Native Americans—to say nothing of the impossibly negative status such a view offers for individuals such as the Métis LeClaire or Jane Johnston Schoolcraft. Again, I instead argue that we should read *Life of Black Hawk* as a counter-captivity narrative in order to highlight—and not obscure—its cross-cultural compositional history. Without finding willing collaborative partners, Black Hawk would have been unable to create a text.

So how did the relationship between Black Hawk and his publishers come about, and was Black Hawk really the one who came up with the idea to publish the book? None of the men responsible for writing *Life of Black Hawk* ever wrote or collaborated on another book. This fact, along with the consistent statements by LeClaire and Patterson regarding their techniques of transcription, suggests that Black Hawk himself was both the primary narrator and the instigator of the creation of the book. Nichols, for one, admits that we cannot know exactly how the collaborative arrangement came about, but that it seems at least "equally possible" that the text was indeed Black Hawk's idea (xv). This should not be surprising—during his tour of the East as a captive, Black Hawk witnessed the desire of sympathetic whites to see and hear him. Jackson's scholarly introduction to the text breaks in tone briefly to suggest that a tour such as his surely featured "dozens of well-wishers [who] would hasten to tell him, 'You ought to write all this down and send it in'" (26). Through his prolonged contact with these audiences, Black Hawk likely was able to develop a savvy that allowed him to strike an acceptably resistant tone. He notes in the text that during the tour in the East, "my opinions were asked on different subjects—but for want of a good interpreter, were very seldom given" (87). As Tena L. Helton argues, Black Hawk had become a celebrity and his presence "becomes a 'social text' as a response to and interaction with newspaper renderings of him" (512).

But beyond what he may have been told and seen in the East, upon his return home Black Hawk was expected to be subordinate to Keokuk, and the survivors of the Black Hawk War were to be "dispersed among different villages so that the group would no longer exist" (Nichols xiv). This disturbing news of Black Hawk's loss of influence must have served as strong motivation to search for new ways to express his resistance and shape his legacy. Sweet's article "Masculinity and Self-Performance in the *Life of Black Hawk*" argues that Black Hawk felt "emasculated" and the text was consequently "an attempt to preserve his traditions" (477). Regardless of the primary reason or even the primary instigator of *Life of Black Hawk*, Black Hawk certainly had considerable motivation to tell his story. After all, Catlin's "Black Hawk and Five Other Saukie Prisoners" demonstrates that Sauk leaders worked with whites in an attempt to communicate their views to a large audience. Black Hawk, then, should not be viewed as if he were exploited by Patterson for profit; the opposite may be closer to the truth. By participating in a counter-captivity narrative, Black Hawk was coopting the skill and ability of Patterson and LeClaire for his own ends. The lasting success of *Life of Black Hawk* has ensured not that Patterson and LeClaire would be well-known, but that Black Hawk would be seen as a great leader despite the disastrous losses suffered during the Black Hawk War.

Most critics who have examined the creation of *Life of Black Hawk* have given some attention to the role of Patterson, who was a newspaper editor for most of his life, but they skip over the role of Black Hawk's translator, Antoine LeClaire, who I believe had at least as important a hand in the composition of *Life of Black Hawk* as did Patterson. A fascinating figure in his own right, LeClaire was born in what is now Michigan to a Potawatomi mother and French-Canadian father in 1797, and was presumably familiar with the practices of captivity, as he spent his life on the frontier. As captivity narratives were a popular genre, and pamphlet-length narratives were often published on the frontier, LeClaire also was likely to have read and been familiar with examples of the genre. LeClaire had been an interpreter near Saukenuk from 1818 onward and is someone that Black Hawk and the other Sauks trusted by and large (Nichols xiv). Few interpreters have such a long record of service and respect from both Native Americans and white traders and soldiers. LeClaire actually appears in *Life of Black Hawk* where Black Hawk first finds white settlers occupying Saukenuk during the winter of 1828–29. Unable to communicate with the settlers, Black Hawk journeys to the Rock Island Indian Agency. LeClaire writes a letter explaining Black Hawk's request for the settlers to leave. Black Hawk thus has a longstanding collaborative relationship with LeClaire, which is yet another reason why he may have contacted LeClaire with the idea of writing his story.

Despite the general willingness of Black Hawk to work with LeClaire, their relationship was not without contention. For example, in the summer of 1829 Black Hawk talked to Neopope about conducting a string of killings (one of the Sauk leaders who later was taken captive with him at the end of the Black Hawk War). At the top of their list was the trader George Davenport—who had apparently purchased three thousand acres of land, including Saukenuk. They also considered killing the Indian agent, translator LeClaire, and others including Keokuck. Black Hawk notes that these were "the principle persons to blame for endeavoring to remove us" (54). Later, he also states that the interpreter—presumably LeClaire—"had been equally as bad [as the agent] in trying to persuade us to leave our village" (58).

Nonetheless, at the treaty signing that ended the Black Hawk War, it was at the request of the Meskwaki and Sauk delegates that LeClaire was singled out to receive two sections of land. This highly valuable property that rapidly made him an exceptionally wealthy trader and, consequently, a founder of Davenport, Iowa. One of the sections may have been intended for Marguerite LeClaire, Antoine's wife (Sherfy 121–22).

Perhaps the most important factor in LeClaire's longstanding connections and involvement with Black Hawk and the other Native Americans around Rock Island was his wife, Marguerite. Marguerite's mother was a Sauk woman, and Marguerite married LeClaire in 1820 (Sherfy 120). For Sauks, kinship is very important, and all indications are that Marguerite regularly offered hospitality to Sauks and Meskwakis. Indeed, for decades after the Black Hawk War, the land that had been granted to the LeClaires hosted annual gatherings of Sauks and Meskwakis (Sherfy 122). Just as the Johnston family enjoyed great influence in Sault Ste. Marie during the same time period due to their Métis connections, the LeClaire family became one of the wealthiest and most influential families in the rapidly growing town of Davenport.[5]

The counter-captivity narrative

The idea of a counter-captivity narrative relies upon the notion that Native American writers were able to adapt the narratives of white captives to their own ends, modifying the written genre along the way. While many captivity narratives were written in an attempt to justify the usurpation of Native American lands through the reinforcement of cultural assumptions of white superiority, *Life of Black Hawk* offers a parallel argument. Black Hawk argues for the superiority of the Sauk Nation, justifying his armed defense of their homelands and reflecting upon what has been lost. For example, Black Hawk explains that he chose to ally himself with the British in 1812 because he "*had not discovered one good trait in the character of the Americans that had come to the country!* They made *fair promises,* but *never fulfilled them!* Whilst the *British* made but few—but we could always *rely upon their word*!" (21; emphasis in original). Patterson may well have allowed this argument to remain due to the formulaic nature of its appeal, which, as I argue, has many parallels with captivity narratives. Although the Americans here are the ones repeatedly denigrated, while the Sauks are described as superior, this pattern does reflect the conventions of the captivity narrative. Although shocking in content, Black Hawk's rhetoric was not represented as an ongoing threat, and so the anger displayed is never frightening to readers. Later, Black Hawk defines himself culturally in response to a request that he move westward, and a challenge of his authority as a leader. Black Hawk angrily replies, "'I am a *Sac!* My forefather was a SAC! And all the nations call me a Sac!!'" (59). To a readership that demanded cultural loyalty to a person's original community, Black Hawk's unflinching statements of tribal allegiance offered a familiar argument.

Another central trope of captivity narratives is the defense of the reputation of a captive by addressing rumors. This appeal is common to captivity narratives because captives were often viewed with suspicion, especially if they had lived with the "Other" for an extended period. Preserving one's cultural identity and avoiding assimilation while captive obviously becomes more difficult during protracted captivity. Black Hawk seems to have required a related type of vouching, as he was accused widely in the press of abandoning his people and culture in a quest for self-preservation. These claims were often based in the fact that Black Hawk was not captured in the so-called Battle of Bad Axe. The certification of authenticity by the translator appears on the first page of Black Hawk's narrative, and LeClaire writes that

> Black Hawk, did … express a great desire to have a History of his life written and published, in order '[that] the people of the United States … might know the causes that impelled him to act as he had done, and the *principles* by which he was governed' (5).

Similar reasoning appears in Patterson's 1834 advertisement in the front matter of the text, where the editor denies any responsibility for the views of Black Hawk, and even claims that he "neither asks, nor expects, any fame for his [Patterson's] services as an amanuensis" (4). In the text itself, Black Hawk concludes by directly confronting rumors: "Before I take leave of the public, I must contradict the story of some *village criers,* who (I have been told,) accuse me of 'having murdered women and children among the whites!' This assertion is *false!* I never did, nor have I any knowledge that any of my nation ever killed a white woman or child" (88; emphasis in original). Precisely because of this, Brumble argues that *Life of Black Hawk* (and *Geronimo's Story of His Life*) is a text motivated by "autobiographical self-vindication" (38); Helton summarizes that "Black Hawk wanted to set the record straight" (512). In other words, Black Hawk was at least partially interested in collaboration in the service of justifying his actions publicly.

Indeed, this strategy seems to have worked, as the authenticity of the narrative and the accuracy of Black Hawk's claims have never been seriously impugned. Only much later, in the 1882 edition that Patterson published with new material, was the factual basis of the text questioned. Later editions of Black Hawk's narrative reverted to the text of the 1834 edition, and this practice continues with the most recent editions. This trajectory is very similar to the publication history of other nineteenth-century captivity narratives, which often began with a relatively high degree of authenticity, then later on—often after the captive had died—were republished with additional fictionalized episodes, only much later to be restored to the original edition.

Even as early as 1833, Patterson attempted to make the text further conform to the expectations of a white audience familiar with captivity narratives. Patterson may even have invented Black Hawk's dedication of the book to his captor, General Atkinson—a dedication that presents the book as the work of a defeated prisoner, rather than featuring the defiant tone that is so prominent in the rest of the text. On the other hand, Black Hawk may have wished such a dedication to serve as an opening salvo declaring himself as equal to Atkinson, and only made prisoner due to what Black Hawk calls "the changes of fortune, and vicissitudes of war" (7).

Additions to the front matter have parallels in many captivity narratives. For example, Mary Rowlandson's narrative was first published with a long introduction by Increase Mather, in addition to a sermon by Rowlandson's minister husband. And just as Rowlandson would have been unable to publish her life story if not for the material fact of her redemption to civilization, so too would Black Hawk have been unable to publish his narrative—especially because it is so critical of whites—if not for the treaty that concluded the Black Hawk War and opened much of eastern Iowa to white settlement. If many colonial captivity narratives were seen to model upright behavior and to affirm the lasting strength of cultural values, so too can *Life of Black Hawk* be seen as a model for proper behavior. That is to say, Black Hawk's narrative was seen by its intended audience of white Americans as extolling the virtues of a leader who refused to relinquish his cultural beliefs even in the face of great trials. For example, Black Hawk describes the crane dance, which was a way that men would court women. He notes that the first year of marriage is a sort of trial in which the couple "ascertain whether they can agree with each other, and can be happy—if not, they part, and each looks out again. If we were to live together and disagree, we should be as foolish as the whites!" (43). Such examples appear throughout *Life of Black Hawk*. Near the conclusion of the prison tour of the East, Black Hawk notes that he was shown "the fire-works, which was quite an agreeable entertainment—but to the *whites* who witnessed it, less *magnificent* than the sight of one of our large *prairies* would be when on fire" (84). Black Hawk here grudgingly admits to some awe at the spectacles presented to him in cities like New York, yet manages to introduce a high degree of cultural relativism by pointing out that whites are similarly impressed when shown prairie fires.

Black Hawk's defiant tone is established early and often in the text. He consistently details abuse as originating from the white settlers. Yet he regularly points out his own decency and temperateness:

> Bad, and cruel, as our people were treated by the whites, not one of them was hurt or molested by any of my band. I hope this will prove that we are a peaceable people—having permitted ten men to take possession of our corn-fields; prevent us from planting corn; burn our lodges; ill-treat our women; and *beat to death* our men, without offering resistance to their barbarous cruelties. This is a lesson worthy for the white man to learn: to use forbearance when injured. (52)

Black Hawk thus argues that his culture is the one that is superior, notably using the word "barbarous" to describe white settlers and "peaceable" for the Sauks.

Another parallel between Black Hawk and the authors of other captivity narratives is his high degree of acculturation. Black Hawk's entire life was spent on or near contested borders, and he was acquainted with both British and American traders and pioneers. Many white captives found themselves identifying with their captors; the same process can be seen in *Life of Black Hawk*. For example, near the end of the text appear many whites whom Black Hawk apparently appreciated: "I cannot omit the name of my old friend CROOKS, of the American Fur Company" (84); and, "I have a good opinion of the American war chiefs, generally, with whom I am acquainted" (86). Black Hawk continually emphasizes his equivalent stature to these powerful whites. Perhaps even more telling, many white Iowa settlers who lived near Black Hawk's cabin in what is now southern Iowa after the publication of the book also recorded friendly visits with him. Sherfy notes that Black Hawk

> frequently visited his [white] neighbors [and] often entertained them—and even strangers—at his own home as well. After the war, Black Hawk became a curiosity and … attracted many visitors. Some went so far as to accuse Black Hawk as being the center of a cult (298).

Black Hawk's keen sense of audience and personal charisma must have been useful while composing *Life of Black Hawk* with Patterson and LeClaire.

Yet Black Hawk was not simply writing a book in order to clear his own name and demonstrate that he was not at fault for the loss of Saukenuk. He understood that without resistance to white encroachment, Native Americans would be pushed westward until they had no place left. He states that during his return to Iowa from the Eastern prison tour,

> I discovered a large collection of [white] people in the mining country, on the west side of the river [...]. I was surprised at this, as I had understood from our Great Father [Andrew Jackson], that the Mississippi was to be the dividing line between his red and white children [...]. I have since found the country much settled by the whites further down and near to our people, on the west side of the river. I am very much afraid, that in a few years, they will begin to drive and abuse our people, as they have formerly done. I may not live to *see* it, but I feel certain the day is not distant. (85; emphasis in original)

Black Hawk here displays a commitment to the Sauk people that lasts beyond the Black Hawk War and suggests another reason he wanted his book published. By invoking his promise from President Jackson that the Mississippi would serve as a barrier to further white encroachment, Black Hawk effectively uses the rhetoric of broken treaties and his authority as a leader to publicize the fact that the battle over land rights is not over.

Life of Black Hawk is in many ways similar in form, textual history, and authorial intentions to captivity narratives. Rather than reading it as an "anomaly" that initiated the genre of American Indian autobiography as Snelling did in 1835, we would do well to understand this cross-cultural collaborative book in the dual contexts of compositional collaboration and the genre of the captivity narrative. Black Hawk, Patterson, and LeClaire were able to exploit existing generic conventions to create a lasting contribution to American literature. As a counter-captivity narrative, *Life of Black Hawk* stands as a contextually grounded text that effectively resists romanticized ideas of Native Americans as well as the policies of Indian removal.

Black Hawk State Historic Site

Finally, *Life of Black Hawk* has undeniably had a lasting impact on the locale Black Hawk knew as Saukenuk. Here is Black Hawk's statement of allegiance to place: "For this spot I felt a sacred reverence, and never could consent to leave it, without being forced therefrom" (56). As readers knew, however, he was indeed forced to leave. Only a decade after the book was published, Margaret Fuller visited the Rock River, and she describes following Black Hawk's trail in *Summer on the Lakes, in 1843*: "How fair the scene through which it led! How could they let themselves be conquered, with such a country to fight for!" (31). Despite Fuller's mischaracterization of Black Hawk's willingness to fight for his homeland, she was an early tourist attracted to the area by the events of the Black Hawk War and by *Life of Black Hawk* itself.

Two hundred acres on a bluff known as Black Hawk's Watch Tower overlook the Rock River, just upstream from the Mississippi, in what is now Rock Island, Illinois. Today, Black Hawk State Historic Site, which is located there, features a museum of Native American history, a memorial statue of Black Hawk, an environmental education center named after Black Hawk's wife, Singing Bird, and several miles of historic trails through the white oak forest.

Black Hawk's narrative was instrumental in ensuring that the main village area of Saukenuk was not divided into small town lots. George Davenport, a wealthy settler who worked on the frontier successively in the US Army, as a trader, and as an Indian agent, purchased the land, to Black Hawk's great distress, in 1830. The events of the Black Hawk War, which included the exhumation of several Indian graves by soldiers at Saukenuk, confirmed the worst of Black Hawk's fears. Patterson was friends with the Davenports, and he dedicated the expanded 1882 edition of Black Hawk's narrative to George's son, Bailey, who was by then owner of the two hundred acres in question. In the 1882 edition, in a new chapter following the main body of the narrative, Patterson claims that Black Hawk himself directly urged George Davenport to hold onto the land. George replied to Black Hawk that he could visit the graves of his ancestors freely, and that the graves would be protected (143). Patterson's slightly discordant conclusion is to note that "for more than half a century, [this land] has been the admiration of many thousands of people" (143). Indeed, in this chapter, entitled "The Black Hawk Tower," he advertises that Bailey Davenport constructed a branch off the train intended for sightseers to visit the high bluffs that had become known as Black Hawk's Watch Tower. Earlier in the text, one of the major additions to the text—an addition that was intended to be part of the main body of the narrative itself—Black Hawk's Watch Tower was the scene of Patterson's invented story of the star-crossed Sioux and Sauk lovers.

Perhaps the success of Patterson's new edition of Black Hawk's narrative provided incentive for Bailey Davenport to develop Watch Tower Park, which soon included an open pavilion for picnickers. Tourism only picked up from there, as Black Hawk's text and image were used in local newspaper advertisements for Watch Tower Park, which was rapidly developed in the 1890s into an amusement park. The first "Shoot the Chute" ride was built here (a toboggan on rails plunging down the bluff into the river), and later a large roller coaster was added. Once the street cars were electrified, tens of thousands of people visited annually. A massive structure that featured a restaurant, dance hall, and ice cream parlor was built to replace an earlier inn that burned in 1896. This building stood for the next twenty years, until it too succumbed to a fire in 1916 ("History"). Still, even at the height of the area's use as an amusement park, the specter of Black Hawk was regularly invoked, along with the spot's natural beauty. In 1896, two hundred and fifty veterans of various US wars met at the Watch Tower (presumably all were white veterans). Gen. Sanders of Rock Island "spoke of the beautiful scenery about Black Hawk's Watch Tower, on which he first stood fifty-one years ago, and beheld the surrounding country, which then, as now, after extended travels, he considered the most beautiful he ever saw" (Anon., "Old Soldiers" 8).

However, the emergence of private automobiles caused electric streetcars to fall into disuse, making Watch Tower Park a commercial failure. The state of Illinois purchased the land and designated it Black Hawk State Park in 1927, demolishing the amusement park. Contemporary arguments for the new state park emphasized the natural features of the landscape. A 1928 newspaper story described the new park this way:

> [T]he place is unsurpassed in beauty. The scenic charm of the river, the woods—most of the tract is wooded—rich in bird life and wild flowers, the tower itself with its commanding view of the surrounding country, sandstone ledges along the river, are natural features which appeal to the desire for the beautiful and the interesting. … Now the descendents of those who fought can enjoy the beauty, can laugh and chat, in the very place where their ancestors gave up their lives for its possession.
>
> ("Black Hawk Park" 8)

This argument emphasizes the natural features of what was once Saukenuk. Interestingly, the importance of the land to the Sauk and Meskwaki is also referenced.

A similar article on the state park was published in 1930, making the connections to land preservation and Native American heritage even more explicit:

> It is as a memorial to the red man whose feet formed these paths that this land has been set aside as a state park. The natural features around Black Hawk state park have been preserved to a remarkable degree, for none of it ever has been under the white man's plow. The woodland is largely white oak but altogether there have been catalogued 45 varieties of trees and a dozen or more shrubs and vines growing on the grounds.
>
> (Patton 9)

While a few hundred acres of protected land may not sound significant, Illinois and Iowa rank 49th and 50th, respectively, in terms of presettlement vegetation remaining. The official web site of Black Hawk State Historic Site notes that "[n]early 175 species of birds can be observed during the year. … More than 30 wildflower species, including orchids (showy orchids), bloom in April and May" ("Natural Features"). Despite the emphasis on nature at Black Hawk State Historic Site, particularly through the Singing Bird Nature Center, Native American history is the primary focus of the Historic Site. The picnics of US Army veterans gave way fairly quickly to Native American powwows. Or perhaps more accurately, the land was shared across cultures, a practice that continues today at what has since been rededicated a State Historic Site. By changing the status of the area from "State Park" to "State Historic Site," I believe that Illinois better recognizes what Saukenuk meant to Black Hawk—a diverse natural area that was actively inhabited by the Sauk.

With many other sites in Iowa, Illinois, and southern Wisconsin preserved because they were sites of the Black Hawk War, or are now simply named after Black Hawk, the influence of *Life of Black Hawk* can hardly be overstated. According to the US Geological Survey, hundreds of places are named after Black Hawk, including Iowa's Black Hawk County and Black Hawk State Park, as well as Wisconsin's Black Hawk Ridge and Black Hawk Lake. While the Black Hawk War may have ignited national interest in Black Hawk, *Life of Black Hawk* has ensured that Black Hawk's motivations are well-remembered. As mentioned earlier, sites named after Black Hawk are often advertised with brief excerpts from *Life of Black Hawk* that describe the landscape. Michael Sherfy suggests that memorializing Black Hawk is above all motivated by a desired connection to place: "[a]s a Native person who had resisted removal from his homeland, Black Hawk was easily conflated with the land itself" (302). Although the conflation of people with nature is problematic, white residents of Illinois, Iowa, and Wisconsin (both the original settlers as well as later generations) found that by claiming a connection to Black Hawk, they were claiming that the natural environment they lived in was historically significant and worth preserving. Too often, I believe, white Midwesterners think of their homes as places where nothing of importance happened; Black Hawk's legacy suggests otherwise.

I've examined only one important strain of the environmental legacy of *Life of Black Hawk*, though, of course, Black Hawk was also interested in creating a legacy for the Sauk people, which is a related goal. The text focused on the center of the Sauk nation, and upon the white encroachment that set off a war. The refiguration of Saukenuk in the form of Black Hawk State Historic Site, together with the hybrid uses to which the site is put by diverse communities, suggests that Black Hawk's legacy is more than simply textual.

Yet, as Kerry Trask concludes, simply memorializing Black Hawk is insufficient. Those residents of eastern cities in 1833 who cheered Black Hawk, as well as many people today, are guilty of "confusing sympathy for contrition, idolization for justice, and seeking easy absolution" (Trask 308). This romantic view has too often led to a simplistic legacy typified by sports teams like the Chicago Black Hawks of the National Hockey League or the Black Hawk military helicopter. Such memorialization can result in names that have little meaning contextually. For example, the Atlanta Hawks basketball team was known as the Tri-Cities Blackhawks before the team was moved to Georgia.

By reading *Life of Black Hawk* as an autobiography and attempting to remove Patterson and LeClaire from the book's textual history, I believe we remain on the sidelines, applauding an ideal of Black Hawk as the Vanishing Indian. This stereotype neglects the complexity of history and reinforces racial division. We gain a much better understanding of *Life of Black Hawk* if we instead consider it as a cross-cultural collaborative text and as a captivity narrative. Reading *Life of Black Hawk* as a cross-cultural collaborative text reminds us that an active and long-lasting struggle was waged over Indian relocation. Black Hawk's willingness to reach out to a mixed-race translator and a sympathetic white audience, which resulted in the material existence of *Life of Black Hawk*, suggests that we can overcome American mythologies of inevitability and division. Our collective identity is not so easily reconciled as Margaret Fuller uneasily mused in 1844, when she toured along "Black Hawk's Trail" and wrote that the displacement of Indians by white settlers was "inevitable, fatal; we must not complain, but look forward to a good result" (29).

Black Hawk's long life illustrates a history of contact between whites, Native Americans, and mixed-race people—not all turns of which were disastrous for all Native Americans. Reading *Life of Black Hawk* as a captivity narrative is equally important. By 1830, the captivity narrative had become established as one of the most authoritative genres to challenge the supposed superiority of white civilization. Lee Clark Mitchell documents how many writers, scientists, and politicians "began in the 1820s to trouble over the alarming rate of recidivism among redeemed white captives" (245), noting that captivity narratives were critical texts that argued against a "legacy of racial stereotypes: the good and bad Indian, the noble savage and unconscionable devil" (220). *Life of Black Hawk*, written immediately after an extended captivity, remains today as a masterful challenge to cultural stereotypes. By joining Black Hawk and his collaborators along the Watch Tower and reading his narrative as a counter-captivity narrative, we can understand the intended legacy of resistance, and later, renewal.

Notes

1. All citations are from: Black Hawk. *Black Hawk's Autobiography*. 1833. Ed. Roger L. Nichols. Ames: Iowa State University Press, 1999.
2. Black Hawk, whose full name was "Black Sparrow Hawk" was unlikely to have been named after the common black hawk (*Buteogallus anthracinus*), whose range in the US is primarily limited to the Southwest, particularly Arizona. The bird known commonly as the black sparrowhawk (*Accipiter melanoleucus*) is native to southern Africa. Based on paintings of Black Hawk holding the fan created from his namesake, with the obvious white band for a tail pattern, I believe the most likely candidate is a rare dark morph of the otherwise widespread

broad-winged hawk (*Buteo platypterus*). The only remaining North American raptor with such a tail pattern is the zone-tailed hawk (*Buteo albonotatus*), which, like the common black hawk, is unlikely to have ventured to the Upper Mississippi from the Southwest even as an item of trade.

3. Along with Rock Island and Moline on the Illinois side of the Mississippi, and Bettendorf in Iowa, Davenport is one of the so-called Quad Cities. Until 1948, Bettendorf had not been included and the metropolitan area was called the Tri-Cities.
4. The Hall sisters gave their testimony in 1867, though they were taken captive in 1832. Stevens prints the entire transcript of their testimony, as well as their brother's, on pages 149–57. The sensationalized account of their captivity printed in 1832 is William P. Edwards, *Narrative of the Capture and Providential Escape of Misses Frances and Almira Hall* (1833).
5. Davenport has almost exactly 100,000 residents today.

5 Communitist narratives of exile and restoration

The captivity narratives I study in this book create space for an overdue renewal of landscape for a range of communities. These collaborative texts, first published in the 1820s and 1830s, make ongoing local and national arguments for justice based on universal human rights, bearing witness to the ongoing inhabitation and use of specific places. While I argue in each chapter of this book for ways in which these specific captivity narratives make the case for site-specific inhabitation, captivity's flip side is migration and exile. Rather than being paradoxical, it is a strength that the categories of home, community, and wilderness are changeable and, sometimes, interchangeable. In addition to staking claims on specific places, these texts materially recognize that we all live on contested grounds.

In this concluding chapter, I start with an overview of reading texts as communal works before undertaking a brief review and re-examination of the captivity narratives that are the focus of earlier chapters. What are the limits of using individual narratives, even ones as richly collaborative and complex as these nineteenth-century captivity narratives? What can be gained by going beyond the individuals whose experiences of captivity were written down by reading texts from the larger perspective of a present-day community? Finally, I end with an intriguing model for ecological and communal restoration, located near sites of the 1862 Dakota Conflict in Minnesota.

I have detailed how each of the captivity narratives I examine in depth earlier are associated with places that have long been preserved by federal, provincial, or state authorities; however, in many more cases, landscapes of captivity have a land use history that reflects radical shifts in usage. In other words, protections were not in place and so active restoration and learning is required in order to bring out features that offer meaning and memory for contemporary communities returning or rebuilding after a long exile.

Jace Weaver coins the term "communitism" in his book *That the People Might Live: Native American Literatures and Native American Community*. Weaver fuses "community" with "activism," noting that writers can act in a corresponding way by "participat[ing] in the healing of the grief and sense of exile felt by Native communities" (xiii). Inspired by Weaver, and even borrowing his book title, Arnold Krupat has recently noted that the "experience of exile, the disruption of ceremonial life, affected the People in innumerable ways, all of them baleful" (*"That the People Might Live"* 8).[1]

One of Krupat's primary texts in his 2012 book is *Life of Black Hawk*. He argues that although Black Hawk's individual losses may seem to be complete, he is rehearsing "a national rather than a personal story" that "tell[s] of what is lost as a form of narrative symbolic action functioning in the interest of its recovery" (*"That the People Might Live"* 109; 110). Elegy, for Black Hawk and other Native American speakers and writers, may primarily serve the interest of a communitist goal rather than a personal one. Losses and recoveries cannot be measured solely on an individual basis. Rather, they are counted on a communal level. Texts such as *Life of Black Hawk*, as well as spoken elegies, "offered mourners consolation so that they might overcome their grief and renew their will to sustain communal life" (Krupat, *"That the People Might Live"* 3). Thus my line of inquiry in this conclusion is slightly rephrased from that of the rest of the book. Rather than reading these captivity narratives as intended *for* a distant audience composed of individuals, what if we read them as the product *of* communities, communities that have rich and ongoing lives and ceremonies?

These rhetorical and cultural considerations are not unique to captivity narratives published in the 1820s and 1830s, of course. For example, Robert Dale Parker shows how writers upheld communal or communitist stories in his introduction to his anthology of early Native American poetry *Changing Is Not Vanishing*. Parker notes how the rhetorical goal of many early writers was to "lock in a cultural memory, to make it impossible to lose the recollection of the land and the lives that were lost and changed during removal" (16). He recovers a poem written just a few years after the 1838–39 Cherokee Trail of Tears by Te-con-ees-kee: "Though far from thee Georgia I roam, / My heart in thy mountain land still has its home" (qtd in Parker 16). Another Cherokee, Ruth Margaret Muskrat, writes of the ongoing remembrance of the land and events in 1922: "In the night they shriek and moan, / In the dark the tall pines moan / As they guard the dismal trail" (qtd. in Parker 17). Even the native forests of Georgia, Muskrat seems to be saying, stand in memory of the loss of the Cherokee.

Contemporary Native American writers continue these communitist themes. Poet Deborah Miranda explains the importance of belonging to her own community of Esselen/Chumash: "… I want these poems to say those words that testify to a miracle, that make song out of quivering air: *Here we are, here we are, here we are*" (xiii). In the poem that shares a title with one of her books, "Indian Cartography," Miranda writes of "a people who are fluid, / fluent in dark water, bodies / long and glinting with sharp-edge jewelry, / and mouths still opening, closing / on the stories of our home" (77). Despite the historical losses—and noting elsewhere that the Esselen are a tribe often cited as being culturally extinct—Miranda emphasizes ongoing life, inhabitation, and action. She links Native American communities to the ecologically and culturally important salmon of the California coastal rivers through the placement of people in water, opening their mouths in mimicry of salmon breathing. "Indian Cartography" places a culture on a map, but more critically, the poem also places the Esselen/Chumash firmly in the present, still actively telling stories.

The preservation of individual places, even when accomplished through collaborative arrangements between individuals, may not tell the entire story. Beyond traditional geographies lies decades and even centuries of exile spurred by forced removal and coerced migration. When we look at texts from a communitist perspective, additional flexibility and hope for restoration can be located. Texts can offer knowledge for communities and writers working to regain cultural knowledge and land restoration. For example, Joseph Bruchac wrote about his identity as a well-known Abenaki writer to Vine Deloria Jr. in 2004, noting that "[w]hatever I know about Abenaki history, culture and language came to me as an adult, but I spent over 40 years in that pursuit" (qtd. in DeLucia, "Placing Joseph Bruchac" 85).[2] This explanation underlines the utility of words as a way to support and renew cultural knowledge.

Lisa Brooks, in her groundbreaking book *The Common Pot: The Recovery of Native Space in the Northeast*, offers an etymology of the Abenaki word *awikhigan*, or book, which she notes is similar to conceptions in the Ojibwe language. "The root word *awigha-* denotes 'to draw,' 'to write,' 'to map'" (xxi). Brooks continues, "The other root of the word for 'book,' *–igan*, denotes an instrument. *Awikhigan* is a tool for image making, for writing, for transmitting an image or idea from one mind to another, over waterways, over time" (xxii). Texts can therefore perform the critical work of linking contemporary members of a community to the experiences and knowledge of past members, ultimately helping to overcome exile. *The Common Pot* "is a mapping of how Native people in the northeast used writing as an instrument to reclaim lands and reconstruct communities" (xxii). This instrumentalist view is useful when studying the captivity narratives in this book as well.

Rather than solely focusing on the conventions of the genre at the moment of publication, we need to look at the older ways of telling that the captives were versed in—and were likely invoking with their editors—even as they were forward-looking for their people. This argument shows that the act of telling these individual stories is not to consign people and places to the past, but to keep them vital and alive for future restoration and for returns from exile. Captivity narratives helped record many individual stories with a diverse array of intended audiences, and even today can offer a rich resource for living communities.

Jemison, Tanner, and Black Hawk as communitist narratives

A Narrative of the Life of Mary Jemison links a range of individuals to the spectacular setting of what is now Letchworth State Park. The book's publication in 1824 was intended to assist Jemison in maintaining her land holdings, which she did for a time; the book also spurred others such as William Pryor Letchworth to preserve and restore the landscape. Yet Jemison's text hinges on her status as a member of the Seneca. She would not have been popularly known as "the White Woman" by her contemporaries if she had returned to white society; she would have simply been a white woman. Jemison's account notes that "it was my choice to stay and spend the remainder of my days with my Indian friends, and live with my family" (120). This formulation places kinship and community at the heart of the narrative, and is consistent with the rest of the text. To further underline her values, the final sentence of the entire narrative actually begins with a conditional clause: "If my family will live happily" (160). This predictive formulation based on community life pushes Jemison's concluding words into the future, making the text communitist insofar as it is working to create the physical space for her family to live. Her words are meant to forestall exile, even as they underscore the importance of communal, kin-based relations.

The choice Jemison makes to remain with her "Indian friends"—her community, her family—and not to return to white society shows how complicated the lives of captives are, and how even communitist actions based on a physical location are subject to contradictory choices. Jemison's move to Buffalo Creek Reservation in 1831 suggests her own exile was self-imposed. Although her liminal status as the White Woman afforded her some legal protection for her substantial land holdings, her chosen community had become ruptured from her chosen place. Jemison, only two years before her death, seems to have chosen exile in the interest of community. Not surprisingly, place continues to evolve over time. The Buffalo Creek Reservation proved to be short-lived, yet in the shape of Letchworth State Park, the Gardeau Flats have been preserved. As mentioned in an earlier chapter, Jemison's remains were later moved back to her long-time home, and powwows have been held at Letchworth State Park.

One fascinating example of how *The Life of Mary Jemison* continues to have a communitist role in the life of the Seneca Nation of Indians can be seen in the work of her descendant G. Peter Jemison. He is an important contemporary Native American artist, who is also the Historic Site Manager of the Ganondagan State Historic Site. G. Peter Jemison explains his knowledge about Mary Jemison through a brief genealogy.

> My father was a seventh generation descendent of Mary Jemison from his mother's side and my father's father was also a Jemison. [...] As a child, I went to a reenactment at Letchworth State Park and that's how I learned the story of Mary Jemison.
>
> (qtd. in Lleweyllan).

With a new education center intended to extend programming at Ganondagan year-round, G. Peter Jemison explains the goals of the center in strikingly communitist language:

> Our goal is to tell the world that we are not a people in the past tense. We live today. We have adapted to the modern world, but we still maintain our language, ceremonies, land base, government, lineages

> and culture. When you're a native person, your story is often told by other people. Here, we tell our own story.
>
> (qtd. in Lleweyllan)

Just as Mary Jemison took her narrative into her own hands to tell her own story to the greatest extent possible, for the greatest community possible, so too does G. Peter Jemison with his paintings, drawing, and videos over several decades as well as with his work at Ganondagan. For example, he created a video in 2009 with his son Brenden. Called *The Mahheakantuk in Focus*, the project "interpreted Henry Hudson's journey up the Hudson River and the meaning behind the 1613 Treaty between the Dutch and the Haudenosaunee" (Ace). G. Peter Jemison's diverse cultural productions can be read and interpreted as national stories, as well as individual stories. The importance of family and culture is woven through his artistic and educational work.

Significantly, *A Narrative of the Captivity and Adventures of John Tanner* concludes in a remarkably similar way to *The Life of Mary Jemison*. The final sentence reads "I have some hope that I may be able to go and make another effort to bring away my daughters" (280). Tanner closes his story with his children, underscoring the desired (re-)continuance of family life. His daughters, he claims, are exiled in the northern woods, far from his new life and work in Sault Ste. Marie. Tanner's social status at the time of his narrative's publication was in some ways even more tenuous than Jemison's, as he had lived a subsistence lifestyle rather than one based primarily on agricultural production. Regardless, Tanner's *Narrative* remains a valuable text for many communities. In her introduction to the Penguin edition of Tanner's *Narrative*, Louise Erdrich annotates the contents of her grandparents' bookshelf in their Turtle Mountain Reservation house: "*The New Testament*, *The Book of Mormon*, and Bishop Baraga's *Ojibway Dictionary*" (xi). The fourth book was *A Narrative of the Captivity and Adventures of John Tanner*; one copy was so heavily read by the Erdrich family that "the binding broke and the pages had to be gathered in a heap, secured with rubber bands" (xi).

As Tanner's *Narrative* is a highly accurate account of Ojibwe lifeways and people during the early nineteenth century, it serves a communitist goal of cultural continuance. As is the case with the captivity narratives of both Jemison and Black Hawk, the lengthy descriptions of how to live—ranging from hunting stories to how to raise children, from how to bury the dead to how to conduct war—are not digressions, but rather are central to the story of being an Ojibwe (or Seneca, or Sauk).

In the conclusion to Chapter 4, I noted that many Midwestern places were named after Black Hawk by whites in search of connection and history. Yet today, the largest Sac (Sauk) and Fox tribe is not located not in the Midwest, but in Oklahoma. Krupat asks us to consider reading *Life of Black Hawk* as a "story for the Sauk people. ... [I]n times to come,

Sauk people might read Black Hawk's words both in English and in Sauk" (*"That the People Might Live"* 116). Indeed, not only is this possible today, it can be demonstrated in the interesting ways that Black Hawk's name and likeness is used by the Sac and Fox nation in Oklahoma. The official Twitter account for the tribe, @sfnsocialmedia, actually uses the head and shoulders of Black Hawk from the 1832 Catlin portrait for its profile image. The name "Black Hawk" is also used in the name of such diverse tribal ventures as the health clinic and a casino (Sac & Fox Nation).

The captivity narratives of earlier chapters have had an environmental and place-based legacy, and it is one that continues to evolve with communities and audiences. Reading them as communitist narratives honors the living legacy of the authors, but in a range of places: Black Hawk in Oklahoma, John Tanner in Turtle Mountain, Mary Jemison in Ganondagan.

Restorationist history

One evening in 2008, Brian Williams closed *NBC Nightly News* with a segment he introduced by calling the Shakopee Mdewakanton Sioux Community (SMSC) "one of the original American communities" ("Shakopee Tribe in Minnesota Using Gambling Money to Restore Their Way of Life"). Roger O'Neil reported on the SMSC's efforts to transform hundreds of acres of farmland into native tallgrass prairies southwest of Minneapolis, Minnesota. Stan Ellison, a leader of the restoration effort for the SMSC, explains the tribe's motivations by saying that "[t]he Dakota culture to a large extent was developed in a certain landscape. A couple of hundred acres in the right place can make a difference" ("Shakopee Tribe"). Today, the lands of the SMSC include more than 550 acres of restored prairie. Shakopee Mdewakanton Sioux Community Chairman Charlie Vig explains that

> [O]ur ancestors' lives were organized around their landscape. They lived in their environment. Their food, fiber, and spiritual life were based on the land on which they lived. Our culture and landscape is intimately intertwined. We have to have the landscape in order to preserve our culture.
>
> (qtd. in Shakopee Mdewakanton Sioux Community).

As noted in the *Nightly News*, these restoration efforts are funded by gaming profits, which have transformed the economic situation for some tribes and left others out.

The importance of prairie restoration for the SMSC is linked to a cultural identity that is linked to the land by both whites and Dakota. For the past 150 years, the forests, wetlands, and prairies of the Upper Midwest have been cut, drained, and plowed for agriculture. Indeed, the official seal of Minnesota (created upon statehood in 1858) actually features a Native American riding a horse away from a farmer with a plow. Although many

Native Americans continue to live in Minnesota, statehood explicitly was secured alongside the removal of Dakota and members of other Native American nations that historically resided in what is now Minnesota.[3] The 1862 Dakota Conflict shattered the cohesiveness of the already severely stressed Dakota community, with large groups going to US states to the west as well as north into Canada. The aftermath included a forced march of 1600 Dakota to an internment camp at Fort Sibley in St. Paul, as well as the largest mass execution in American history—reduced by President Abraham Lincoln's pardons from 303 to 38. I do not wish to suggest that the disruption of the tallgrass prairie ecosystem should be measured alongside the horrific human costs, but rather am arguing here that the destruction of the prairie can be seen as part and parcel of white attempts to destroy the cultural identity of the Dakota.

In *The War on Words: Reading the Dakota Conflict through the Captivity Literature*, Kathryn Zabelle Derounian-Stodola analyzes the competing captivity narratives that were told by a total of twenty-four people, including Dakota stories as well as those of whites. She concludes that

> [A]ll those who originally took part in the Dakota War clashed over a place they called, or wanted to call, home … In fighting their own verbal wars after the fact these narratives attest to survival and continuity through storytelling, and they articulate the search for home (277).

Sarah Wakefield's pro-Dakota narrative, *Six Weeks in the Sioux Teepees*, describes her first look at the land in 1861, when her husband was appointed as a medical doctor at Yellow Medicine: "Although I was nervous, I enjoyed that ride, for a more beautiful sight than that prairie, I have never seen. It was literally covered with flowers of all descriptions; the tall grass was waving in the breeze" (55). The beauty of this new home for Wakefield was soon marginalized by the deprivations she observed amongst the Dakota from lack of food and mistreatment by whites, and then the war and her own extended captivity of the following year. Wakefield lobbied for the pardon of her captors, and her narrative is sympathetic to the Dakota, by and large. However, as the American landscape was radically altered alongside the displacement of Native Americans, one fascinating strategy to address the presence of Native Americans in historical homelands may be seen in the SMSC's prairie restoration efforts.

William Jordan III argues that we can link communities to landscapes (and to each other) through environmental restoration. In *The Sunflower Forest: Ecological Restoration and the New Communion with Nature*, he begins by showing why this is a relatively new idea—and why collaborative processes are critical. "[F]or more than a century, environmentalism has … proven incapable of inhabiting the middle ground where community is achieved as selves confront each other, first to acknowledge and then somehow to transcend the irreconcilable differences between them" (46). Some of the value of restoration work can be seen in the way that it actively includes human interaction with the land and with the past. Jordan considers restoration as "performed or applied history. Burning a prairie, for example, reenacts the burning of the prairies by Indians in pre-contact times and dramatically revives their history, making it part of the consciousness of a community" (Jordan 130). Because the conscious decision to alter a landscape is rooted in action, great potential can be seen in these sorts of ecological restorations.

The SMSC shows one potential model to make changes in the land for the better. Looking at historical maps, narratives, and transcripts can inform changes in relationships between human communities as well as the landscape itself. The community integration of the prairie is clearly part of what the SMSC desired when they began the 500-acre restoration effort in collaboration with the Minnesota Department of Natural Resources and the University of Minnesota.

The collaborative narratives this book examines may have originated in individuals who lived and worked two hundred years ago, but they offer an ongoing path forward for living communities. The unique insights made possible through captivity narratives continue to influence social justice efforts and environmental ethics today.

Notes

1. Krupat borrows the title phrase with permission (*"That the People Might Live"* 9).
2. DeLucia's article is based largely on her examination of Bruchac's papers at Beinecke Rare Book and Manuscript Library at Yale University. As she notes, the collection is being continuously updated.
3. The Dakota are part of the larger Sioux nation.

Appendix 1

Publication history of *The Life of Mary Jemison*

Table A.1 Adaptation of the "Tabulation of Editions of 'The Life of Mary Jemison,'" from James Seaver's *A Narrative of the Life of Mary Jemison, the White Woman of the Genesee*. Ed. Charles Delamater Vail. New York: American Scenic & Historic Preservation Society, 1925. 296–97

Ed.	*Year*	*Place*	*Pages*	*Illus.*	*Page Size*	*Editor*	*Publisher*	*Character of Text*
1st	1824	Canandaiga	189	0	3⅜ × 5½		James D. Bemis	
2nd	1826	Howden	180	0	3¾ × 6		R. Parkin	Reprint of 1st
3rd	1827	London	180	0	3⅝ × 5⅞		Longman, Rees	Reprint of 1st
4th	1834	Buffalo	36	0	4½ × 7½	Unknown	Unknown	Abridgement
5th	1840	Rochester	36	3	5⅜ × 8⅛	Unknown	Miller, Butterfield	Same as 4th
6th	1841	Rochester	36	3	4⅞ × 8	Unknown	G. Cunningham	Same as 5th
7th	1842	Utica	32	3	5⅜ × 7⅞	Unknown	G. Cunningham	Same as 6th
8th	1842	Otley	192	1	3 × 4¾	Unknown	William Walker	1st w/ additions
9th	1842	Batavia	192	0	3⅞ × 5¾	Mix	William Seaver	1st much revised
10th	1842	Batavia	192	0	3⅞ × 5¾	Mix	William Seaver	Same as 9th
11th	1844	Batavia	192	0	3¾ × 6	Mix	William Seaver	Same as 9th
12th	1847	Devon, London	184	0	3⅝ × 5⅞	Mix	S. Thorne, W. Tegg	Same as 9th
13th	1856	New York, Auburn, Rochester	312	5	4¾ × 7¼	Morgan	D. M. Dewey	9th revised
14th	1859	New York	312	5	5 × 7¼	Morgan	C. M. Saxton	Same as 13th
15th	1860	New York	312	5	4⅞ × 7⅜	Morgan	Saxton, Barker	Same as 13th
16th	1877	Buffalo	303	17	4¾ × 7½	Letchworth	Letchworth	13th revised
17th	1880	Buffalo	303	17	4¾ × 7¼	Letchworth	Letchworth	Same as 16th
18th	1898	New York, London	300	21	4⅞ × 7⅜	Letchworth	Letchworth	17th revised
19th	1910	New York, London	305	22	5 × 7½	Letchworth	Letchworth	18th revised
20th	1913	New York, London	305	22	5 × 7½	Letchworth	Amer. Scen. & Hist. Pres. Soc.	Same as 19th
21st	1918	New York	475	41	5⅜ × 8	Vail	Amer. Scen. & Hist. Pres. Soc.	1st restored w/ additions and revision
22nd	1925	New York	483	41	5⅜ × 8	Vail	Amer. Scen. & Hist. Pres. Soc.	21st revised

Appendix 2

Publication history of *A Narrative of the Captivity and Adventures of John Tanner*

(includes all American and foreign editions, cited by the original edition's author and publication listing on WorldCat).

1830:

Tanner, John. *A Narrative of the Captivity and Adventures of John Tanner (U.S. Interpreter at the Saut de Ste. Marie) During Thirty Years Residence Among the Indians in the Interior of North America.* Ed. Edwin James. New York: G. & C. & H. Carvill, 1830.

Tanner, John. *A Narrative of the Captivity and Adventures of John Tanner (U.S. Interpreter at the Saut de Ste. Marie) During Thirty Years Residence Among the Indians in the Interior of North America.* Ed. Edwin James. London: Baldwin & Cradock, 1830.

1835:

Tanner, John. *Mémoires de John Tanner, ou Trente Années Dans les Déserts de l'Amérique du Nord.* Trans. Ernest de Blosseville. Paris: Arthus Bertrand, 1835.

1840:

Tanner, John. *Des Kentuckier's John Tanner Denkwürdigkeiten über seinen dreissigjährigrn Aufenthalt unter den Indianern Nord-Amerika's.* Trans. Karl Andree. Leipzig: W. Engelmann, 1840.

1883:

Tanner, John. *Grey Hawk: Life and Adventures Among the Red Indians.* Ed. James Macaulay. Philadelphia: J. B. Lippincott Co., 1883.

1909:

Macaulay, James and John Tanner. *Grey Hawk: Life and Adventures Among the Red Indians.* London: H. Frowde, Hodder and Stoughton, 1909.

1940:

Tanner, James. *An Indian Captivity (1789–1822). John Tanner's narrative of his captivity among the Ottawa and Ojibwe Indians.* Ed. Edwin James. San Francisco: California State Library, 1940.

1956:
Tanner, John. *A Narrative of the Captivity and Adventures of John Tanner (U.S. Interpreter at the Saut de Ste. Marie) During Thirty Years Residence Among the Indians in the Interior of North America.* Ed. Edwin James. Minneapolis: Ross & Haines, Inc., 1956.

1963:
Tanner, John and Edwin James. *Rasskaz o pokhishchenii I prikliucheniiakh Dzhona Tennera: perevodchika na sluzhbe SShA v So-Sent-Mari v techenie tridtsatiletnego prebyvaniia sredi indeitsev v glubine Severnoi Ameriki.* Trans. Unknown. Moscow: Izd-vo inostrannoi lit-ry, 1963.

1968:
Tanner, John. *Dreissig Jahre unter den Indianern. Leben und Abenteuer des John Tanner.* Ed. Edwin James. Trans. Eva Lips. Weimar: Kiepenheuer, 1968.

1975:
Tanner, John and Edwin James. *A Narrative of the Captivity and Adventures of John Tanner.* New York: Garland, 1975.

1981:
Tanner, John. *Dreissig Jahre unter den Indianern Nordamerikas.* Trans. Karl Andree. Munich: L. Borowsky, 1981.

1983:
Tanner, John and Edwin James. *Trente ans de captivité: chez les Indiens Ojibwe.* Trans. Pierrette Desy. Paris: Payot, 1983.

1994:
Tanner, John. *The Falcon: a narrative of the captivity and adventures of John Tanner.* Ed. Louise Erdrich. New York: Penguin Books, 1994.

2000:
Tanner, John. *The Falcon: a narrative of the captivity and adventures of John Tanner.* Ed. Louise Erdrich. New York: Penguin Books, 2000.

2003:
Tanner, John. *The Falcon: a narrative of the captivity and adventures of John Tanner.* Ed. Louise Erdrich. New York: Penguin Books, 2003.

2009:
Tanner, John with Edwin James. *A Narrative of John Tanner "the Falcon": His Indian Captivity and Adventures.* Ed. Charles Daudert. Kalamazoo, MI: Hansa-Hewlett, 2009.

Appendix 3

Publication history of *Life of Black Hawk*

(includes all American and foreign editions, cited by the original edition's author and publication listing on WorldCat).

1833:
Black Hawk. *Life of Ma-ka-tai-me-she-kia-kiak, or Black Hawk, embracing the tradition of his nation, Indian wars in which he has been engaged, cause of joining the British in their late war with America, and its history, description of the Rock-River village, manners and customs, encroachments by the whites, contrary to treaty, removal from his village in 1831. With an account of the cause and general history of the late war, his surrender and confinement at Jefferson barracks and travels through the United States.* Cincinnati: J. B. Patterson, 1833.

1834:
Black Hawk. *Life of Ma-Ka-Tai-Me-She-Kia-Kiak, or Black Hawk, embracing the tradition of his nation, Indian wars in which he has been engaged, cause of joining the British in their late war with America, and its history, description of the Rock-River village, manners and customs, encroachments by the whites, contrary to treaty, removal from his village in 1831. With an account of the cause and general history of the late war, his surrender and confinement at Jefferson Barracks, and travels through the United States.* Ed. J. B. Patterson. Boston: Russell, Odiorne & Metcalf, 1834. Published concurrently by four other presses. New York: Monson Bancroft; Philadelphia: Marshall, Clark & Co.; Baltimore: Jos. Jewett; and Mobile: Sidney Smith.

1836:
Black Hawk. *Life of Ma-Ka-Tai-Me-She-Kia-Kiak, or Black Hawk, embracing the tradition of his nation, Indian wars in which he has been engaged, cause of joining the British in their late war with America, and its history, description of the Rock-River village, manners and customs, encroachments by the whites, contrary to treaty, removal from his village in 1831. With an account of the cause and general history of the late war, his surrender and confinement at Jefferson Barracks, and travels through the United States.* Ed. J. B. Patterson. London: R. J. Kennett, 1836.

1842:
Black Hawk. *Life of Ma-Ka-Tai-Me-She-Kia-Kiak, or Black Hawk, embracing the tradition of his nation, Indian wars in which he has been engaged, cause of joining the British in their late war with America, and its history, description of the Rock-River village, manners and customs, encroachments by the whites, contrary to treaty, removal from his village in 1831. With an account of the cause and general history of the late war, his surrender and confinement at Jefferson Barracks, and travels through the United States.* Ed. J. B. Patterson. Cooperstown, N.Y.: H. & E. Phinney, 1842.

1845:
Black Hawk. *Life of Ma-Ka-Tai-Me-She-Kia-Kiak, or Black Hawk*. Boston: T. Abbot, 1845.

1882:
Black Hawk and J. B. Patterson. *Autobiography of Ma-Ka-Tai-Me-She-Kia-Kiak, or Black Hawk, Embracing the traditions of his nation, various wars in which he has been engaged, and his account of the cause and general history of the Black Hawk War of 1832, his surrender, and travels through the United States. Also, Life, Death, and Burial of the old chief, together with a history of the Black Hawk War*. Interpreter Antoine LeClaire. Ed. and amanuensis J. B. Patterson. St. Louis: Continental Printing Co., 1882. The expanded edition with Patterson's additions.

1912:
Black Hawk. *Life of Black Hawk, or Ma-Ka-Tai-Me-She-Kia-Kiak*. Ed. James D. Rishell. Rock Island, IL: American Publishing Co., 1912.

1916:
Black Hawk. *Life of Black Hawk, or Ma-Ka-Tai-Me-She-Kia-Kiak*. Ed. Milo Milton Quaife. Lakeside Classics. Chicago: R. R. Donnelley & Sons, 1916. Reprint of 1834 Boston edition.

1932:
Black Hawk. *Life of Black Hawk, or Ma-Ka-Tai-Me-She-Kia-Kiak*. Ed. J. B. Patterson. Trans. Antoine LeClaire. Iowa City: Iowa State Historical Society, 1932. Published on the 100th anniversary of the signing of the treaty that ended the Black Hawk War.

1955:
Black Hawk. *Black Hawk, an Autobiography*. Ed. Donald Jackson. Urbana: U of Illinois Press, 1955.

1964:
Black Hawk. *Black Hawk, an Autobiography*. Ed. Donald Jackson. Urbana: U of Illinois Press, 1964.

1973:
Black Hawk. *Black Hawk, an Autobiography*. Ed. Donald Jackson. Urbana: U of Illinois Press, 1973.

1974:
Black Hawk. *Life of Ma-Ka-Tai-Me-She-Kia-Kiak, or Black Hawk*. Ed. J. B. Patterson. Intro. James Dowd. Fairfield, WA: Ye Galleon Press, 1974. Reprint of the 1834 Boston edition.

1989:
O Suti Zhizni: Indeiskie i Eskimosskaia avtobiografii. Trans. Aleksandr Vashchenko. IAkutsk: IAkutskoe Knizhnoe Izd-vo, 1989. Translation of *Life of Black Hawk*, *Black Elk Speaks*, and Nuligak's *Mémoires d'un Esquimau*.

1990:
Black Hawk. *Black Hawk, an Autobiography*. Ed. Donald Jackson. Prairie State Books. Urbana: U of Illinois Press, 1990.

1994:
Black Hawk. *Life of Black Hawk, or Ma-Ka-Tai-Me-She-Kia-Kiak*. Ed. Milo Milton Quaife. New York: Dover, 1994. Reprint of the 1916 Chicago edition.

1999:
Black Hawk. *Black Hawk's Autobiography*. Ed. Roger L. Nichols. Ames: Iowa State University Press, 1999. Reprint of 1834 Boston edition.

2002:
Black Hawk. *Autobiography of Ma-Ka-Tai-Me-She-Kia-Kiak, or Black Hawk*. Ed. J. B. Patterson. Bowie, MD: Heritage Books, 2002. Reprint of the 1882 edition.

2008:
Black Hawk. *Life of Black Hawk, or Ma-Ka-Tai-Me-She-Kia-Kiak*. Ed. J. Gerald Kennedy. New York: Penguin Books, 2008.

Bibliography

Primary texts

"Agreement with the Seneca, 1797." *Indian Affairs: Laws and Treaties*. Vol. II. Ed. Charles J. Kappler. Washington, D.C.: Government Printing Office, 1904. 1027–30.

Anonymous. "American Literature." *The Western Monthly Magazine, and Literary Journal* 1.1 (Jan. 1833): 5–9.

Anonymous. "Black Hawk Park is Rich in Tradition and History." *The Decatur Daily Review*. (July 15, 1928): 8.

Anonymous. "The Indian Captive Reclaimed." *The Philadelphia Register and National Recorder* 1.7 (Feb. 13, 1819): 127.

Anonymous. "John Tanner." *Christian Watchman* 9.34 (Aug. 22, 1828): 136.

Anonymous. "Muck-A-Tay Mich-E-Kaw-Kiak, the Black Hawk." *The New York Mirror: A Weekly Gazette of Literature and the Fine Arts* 11.2 (July 13, 1833): 9–10.

Anonymous. "Old Soldiers in Reunion Today at the Watch Tower." *Davenport Daily Leader*. (August 28, 1896): 8.

Anonymous. "Phrenological Developments and Character of the Celebrated Indian Chief and Warrior, Black Hawk; with Cuts." *American Phrenological Journal* 1.2 (Nov. 1, 1838): 51–62.

Anonymous Book Review. "Art. XVII.–Account of an expedition from Pittsburgh to the Rocky Mountains …" *The North American Review* 7.2 (Apr. 1823): 242–69.

Anonymous Book Review. "A Narrative of the Captivity and Adventures of John Tanner …" *The American Quarterly Review* 8.15 (Sept. 1830): 108–34.

Anonymous Book Review. "Tanner's Captivity Among the Indians." *The Albion: A Journal of News, Politics and Literature* 9.21 (Oct. 30, 1830): 163.

American Scenic and Historical Preservation Society. "Appendix A, Dedication of the Statue of Mary Jemison, the White Woman of the Genesee, At Letchworth Park, September 19, 1910." *Sixteenth Annual Report, 1911, of the American Scenic and Historical Preservation Society*. Albany, N.Y.: J. B. Lyon, 1911. 229–259.

Apess, William. *A Son of the Forest*. 1831. *On Our Own Ground: The Complete Writings of William Apess, a Pequot*. Ed. Barry O'Connell. Amherst: U of Massachusetts P, 1992.

Black Hawk. *Black Hawk's Autobiography*. 1833. Ed. Roger L. Nichols. Ames: Iowa State UP, 1999.

Bryant, William Clement. "Chapter II: Mary Jemison's Indian Name." 1877. *A Narrative of the Life of Mary Jemison, the White Woman of the Genesee*. By James E. Seaver. Ed. Charles Delamater Vail. 22nd ed. New York: American Scenic and Historic Preservation Society, 1925. 198–207.

Bryant, William Cullen. "The Maiden's Sorrow." *Graham's Lady's and Gentleman's Magazine* 21.2 (August 1842): 64.

Cabeza de Vaca, Álvar Núñez. *The Narrative of Cabeza de Vaca*. 1542. Trans. Rolena Adorno and Patrick Charles Pautz. Lincoln: U of Nebraska P, 2003.

Catlin, George. *Letters and Notes on the North American Indians*. New York: Clarkson N. Potter, 1975.

Cooper, James Fenimore. *The Last of the Mohicans*. 1826. New York: Penguin, 1986.

Cooper, Susan Fenimore. *Rural Hours*. 1850. Eds. Rochelle Johnson, and Daniel Patterson. Athens: U of Georgia P, 2005.

Crèvecoeur, J. Hector St. John de. *Letters from an American Farmer*. 1782. Ed. Albert E. Stone. New York: Penguin, 1986.

Edwards, William P. *Narrative of the Capture and Providential Escape of Misses Frances and Almira Hall: Two Respectable Young Women (Sisters) of the Ages of 16 and 18, who were taken Prisoners by the Savages, at a Frontier Settlement, near Indian Creek* ... Publisher unknown, 1833.

Fuller, Margaret. *Summer on the Lakes, in 1843*. 1844. Urbana: U of Illinois P, 1991.

Goodale, Elaine and Dora Read Goodale. *In Berkshire with the Wild Flowers*. New York: G. P. Putnam's Sons, 1879.

Gray, David. "To Glen Iris [Impromptu]." 1876. *Voices of the Glen*. Ed. Henry R. Howland. Adminstrator, William Pryor Letchworth. New York: Knickerbocker P, 1911. 19–20.

———. "The Last Indian Council on the Genesee." *Scribner's Monthly* 14.3 (July 1877): 338–50.

Harmon, Daniel Williams. *A journal of voyages and travels in the interiour of North America, between the 47th and 58th degrees of north latitude, extending from Montreal to the Pacific ocean ... including an account of the principal occurrences, during a residence of nineteen years, in different parts of the country: to which are added, a concise description of the face of the country, its inhabitants ... and considerable specimens of the two languages, most extensively spoken: together with an account of the principal animals, to be found in the forests and prairies of this extensive region*. Andover: Flagg and Gould, 1820. *Sabin Americana*. Gale, Cengage Learning. University of Nevada Reno. 11 August 2009.

Herr, Wm. "Revival Intelligence, For the Western Christian Advocate, Detroit District, Mich. Conference." *Western Christian Advocate* 5.3 (May 11, 1838): 10.

Irving, Washington. *A Tour on the Prairies*. 1835. Norman: U of Oklahoma P, 1956.

Gyles, John. *Memoirs of Odd Adventures, Strange Deliverances, etc. in the Captivity of John Gyles, Esq.; Commander of the Garrison on St. George's River*. Boston: S. Kneeland and T. Green, 1736.

James, Edwin, ed. *Account of an Expedition from Pittsburgh to the Rocky Mountains, Performed in the years 1819 and '20, by order of the Hon. J. C. Calhoun, Sec'y of War: Under the Command of Major Stephen H. Long*. Philadelphia: H. C. Carey and I. Lea, 1823.

———. "Diary and Journal Notes, 1820–1827." 3 vols. bound as one. Lent to the Bancroft Library for microfilming in August, 1956, by the Columbia University Library, New York.

———. "Essay on the Chippewa Language, Read before the American Lyceum, at the third Annual Meeting, in the city of New York, May 3rd, 1833." *American Annals of Education* 3.10 (October 1833): 440–46.

———. "Introduction." John Tanner and Edwin James, M.D. *A Narrative of the Captivity and Adventures of John Tanner (U.S. Interpreter at the Saut de Ste. Marie)*

During Thirty Years Residence Among the Indians in the Interior of North America. 1830. Minneapolis: Ross & Haines, 1956. xvii–xxxiv.

———. "Review of *A Grammar of the Language of the Lenni Lenape or Delaware Indians*. Translated for the American Philosophical Society, from the German Manuscript of the late Rev. David Zeisberger. By Peter Stephen Duponceau. Philadelphia, 1827." *American Quarterly Review* 3.6 (June 1828): 391–422.

Keating, William Hypolitus. *Narrative of an Expedition to the Source of St. Peter's River, Lake Winnepeek, Lake of the Woods, &c., &c., Performed in the Year 1823, by Order of the Hon. J. C. Calhoun, Secretary of War, Under the Command of Stephen H. Long, Major, U.S.T.E.* Philadelphia: H. C. Carey & I. Lea, 1824.

Larcom, Lucy. "Manitou's Garden." *Forrester's Boys' and Girls' Magazine, and Fireside Companion*. Sept. 1, 1854: 80.

Lee, Dr. Chas. A. "The Residence of Tanner, Or the Indian Whiteman." *Dwights American Magazine, and Family Newspaper* 2.26 (July 25, 1846): 388.

Lincoln (Phelps), Almira H. *Familiar Lectures on Botany, Including Practical and Elementary Botany, with Generic and Specific Descriptions of the Most Common Native and Foreign Plants, and a Vocabulary of Botanical Terms, For the Use of Higher Schools and Academies*. Hartford: H. and F. J. Huntington, 1829.

Long, Stephen H. *The Northern Expeditions of Stephen H. Long: The Journals of 1817 and 1823 and Related Documents*. Ed. Lucile M. Kane, June D. Holmquist, and Carolyn Gilman. St. Paul: Minnesota Historical Society P, 1978.

Loudon, Archibald. *A Selection of Some of the Most Interesting Narratives of Outrage Committed by the Indians*. 1808. Vol. 1. Harrisburg, PA: Harrisburg Publishing Co., 1888.

Macaulay, James. *Grey Hawk: Life and Adventures among the Red Indians*. Philadelphia: J. B. Lippincott Co., 1883.

Mix, Ebenezer. "Chapter I" and "Chapter VII." 1842. *A Narrative of the Life of Mary Jemison, the White Woman of the Genesee*. By James E. Seaver. Ed. Charles Delamater Vail. 22nd ed. New York: American Scenic and Historic Preservation Society, 1925. 193–95, 251–63.

Morgan, Lewis Henry. *Houses and House-Life of the American Aborigines*. 1881. Chicago: U of Chicago P, 1965.

———. "Indian geographical names in the territories of the Seneca, Cayugas, Onondagas, Oneidas, and Mohawks, with their corresponding English names and their signification." 1856. *A Narrative of the Life of Mary Jemison, the White Woman of the Genesee*. By James E. Seaver. Ed. Charles Delamater Vail. 22nd ed. New York: American Scenic and Historic Preservation Society, 1925. 264–73.

Patterson, J. B., and Black Hawk. *Autobiography of Ma-Ka-Tai-Me-She-Kia-Kiak, or Black Hawk, Embracing the traditions of his nation, various wars in which he has been engaged, and his account of the cause and general history of the Black Hawk War of 1832, his surrender, and travels through the United States. Also, Life, Death, and Burial of the old chief, together with a history of the Black Hawk War*. Interpreter Antoine LeClaire. Ed. and amanuensis J. B. Patterson. St. Louis: Continental Printing Co., 1882.

Patton, James S. "Black Hawk State Park and Rock Island Arsenal Offer Much of Interest." *Decatur Herald*. July 13, 1930. 9.

Perdue, Chas A. "Domestication of Wild Flowers." *Maine Farmer*. June 25, 1863. 1.

Phelps, Almira Lincoln. *Familiar Lectures on Botany, Practical, Elementary, and Physiological: With an Appendex, Containing Descriptions of the Plants of the*

United States and Exotics, &c., for the Use of Seminaries and Private Students. 5th Ed. Hartford, F. J. Huntington, 1836. *Sabin Americana.* Gale, Cengage Learning. University of Nevada Reno. 16 July 2009.

Plummer, Rachel. *A Narrative of the Capture and Subsequent Sufferings of Mrs. Rachel Plummer, Written by Herself.* 1839. *Held Captive by Indians: Selected Narratives.* 1973. Ed. Richard VanDerBeets. Knoxville: U of Tennessee P, 1994. 333–66.

Rowlandson, Mary. *The Sovereignty and Goodness of God, Together with the Faithfulness of His Promises Displayed, Being a Narrative of the Captivity and Restoration of Mrs. Mary Rowlandson and Related Documents.* 1682. Ed. Neal Salisbury. Boston: Bedford/St. Martin's, 1997.

Sanford, Laura. "Cultivation of Native Orchids." *The Independent: Devoted to the Consideration of Politics, Social and Economic Tendencies, History, Literature, and the Arts.* August 27, 1891: 30.

Schoolcraft, Henry Rowe. *Algic Researches: Indian Tales and Legends.* 1839. Vol. I and II. Baltimore: Clearfield Co., 1992.

———. *The Indian in His Wigwam, or, Characteristics of the Red Race of America: From Original Notes and Manuscripts.* New York: W. H. Graham, 1848.

———. *Information Respecting the History, Condition and Prospects of the Indian Tribes of the United States: Collected and Prepared Under the Direction of the Bureau of Indian Affairs, Per Act of Congress of March 3rd, 1847.* Part IV. Philadelphia: Lippincott, Grambo and Co., 1854.

———. *Personal Memoires of a Residence of Thirty Years with the Indian Tribes on the American Frontiers: With Brief Notices of Passing Events, Facts, and Opinions.* Philadelphia: Lippincott, Grambo and Co., 1851.

Seaver, James E. *A Narrative of the Life of Mary Jemison, the White Woman of the Genesee.* Ed. Charles Delamater Vail. New York: American Scenic & Historic Preservation Society, 1925.

———. *A Narrative of the Life of Mrs. Mary Jemison.* 1824. Ed. June Namias. Norman: U of Oklahoma P, 1993.

Sedgwick, Catharine Maria. *Hope Leslie; Or, Early Times in the Massachusetts.* 1827. New York: Penguin, 1998.

Snelling, William Joseph. "Life of Black Hawk." Review of *Life of Mal-ka-tai-me-she-kia-kiak* [sic] or *Black Hawk, dictated by himself.* By Black Hawk. *North American Review* 40.76 (Jan. 1835): 68–87.

Tanner, John. *The Falcon: A Narrative of the Captivity and Adventures of John Tanner.* New York: Penguin Nature Library, 1994.

———. *A Narrative of the Captivity and Adventures of John Tanner (U.S. Interpreter at the Saut de Ste. Marie) During Thirty Years Residence Among the Indians in the Interior of North America.* Ed. Edwin James, M.D. New York: G. & C. & H. Carvill, 1830.

Territory of Michigan. *Laws of the Territory of Michigan: Embracing the Acts and Resolutions of the Legislative Council for the Years 1830, '31, '32, '33, '34, and '35.* Vol. III. Lansing: W. S. George and Co., 1874.

Thoreau, Henry David. *Walden; or, Life in the Woods.* 1854. Boston: Beacon P, 1997.

Torrey, John. "Descriptions of some new or rare Plants from the Rocky Mountains, collected in July, 1820, by Dr. Edwin James. Read before the Lyceum, Sept. 22, 1823." *Annals of the Lyceum of Natural History of New-York.* Vol. 1. New York: J. Seymour, 1824. 30.

Turner, Frederick Jackson. *Rereading Frederick Jackson Turner: The Significance of the Frontier in American History, and Other Essays*. Ed. John Mack Faragher. New York: Henry Holt, 1994.

Wakefield, Sarah. *Six Weeks in the Sioux Teepees: A Narrative of Indian Captivity*. 1864. Ed. June Namias. Norman: U of Oklahoma P, 1997.

Washington, George. "Instructions to Major General John Sullivan." *The Papers of George Washington, Digital Edition*. Ed. Theodore J. Crackel. Charlottesville: UP of Virginia, 2007.

Webster, William. "An Ornamental Farm." *Moore's Rural New Yorker* 12.11 (March 16, 1861): 87.

Wright, Laura. "Chapter III: Last Hours of the Captive." 1877. *A Narrative of the Life of Mary Jemison, the White Woman of the Genesee*. By James E. Seaver. Ed. Charles Delamater Vail. 22nd ed. New York: American Scenic and Historic Preservation Society, 1925. 208–12.

Wright, Mabel Osgood. *Flowers and Ferns in their Haunts*. New York: Macmillan Co., 1901.

———. *The Friendship of Nature: A New England Chronicle of Birds and Flowers*. 1894. Ed. Daniel J. Philippon. Baltimore: Johns Hopkins UP, 1999.

Secondary texts

Abler, Thomas S. "Protestant Missionaries and Native Culture: Parallel Careers of Asher Wright and Silas T. Rand." *American Indian Quarterly* 16.1 (Winter 1992): 25–37.

Ace, Barry. "G. Peter Jemison." Museum of Contemporary Native Arts. IAIA.edu.

Adamson, Joni. "For the Sake of the Land and All People: Teaching American Indian Literatures from an Environmental Justice Perspective." *Teaching North American Environmental Literature*. Eds. Laird Christensen, Mark C. Long, and Fred Waage. New York: The Modern Language Association, 2008. 194–202.

Adamson, Joni, Mei Mei Evans, and Rachel Stein. "Introduction." *The Environmental Justice Reader: Politics, Poetics, and Pedagogy*. Tucson: U of Arizona P, 2002. 3–14.

Barnett, Louise K. *The Ignoble Savage: American Literary Racism, 1790–1890*. Westport: Greenwood P, 1975.

Beale, Irene A. *William P. Letchworth: A Man for Others*. Geneseo: Chestnut Hill P, 1982.

Benson, Maxine. "Edwin James: Scientist, Linguist, Humanitarian." Diss. U of Colorado, 1968.

———. "Schoolcraft, James, and the 'White Indian.'" *Michigan History* 54.4 (1970): 311–28.

Beidler, Peter G. "The Facts of Fictional Magic: John Tanner as a Source for Louise Erdrich's 'Tracks' and 'The Birchbark House.'" *American Indian Culture & Research Journal* 24.4 (2000): 37–54.

Boardman, Kathleen. "Collaborative Life Narrative." *Encyclopedia of Women's Autobiography*. Vol. 1: A-J. Eds. Victoria Boynton and Jo Malin. Westport, CT: Greenwood P, 2005. 175–77.

Branch, Michael P., ed. *Reading the Roots: American Nature Writing before* Walden. Athens: U of Georgia P, 2004.

Bremer, Richard G. *Indian Agent and Wilderness Scholar: The Life of Henry Rowe Schoolcraft*. Mount Pleasant: Clarke Historical Library and Central Michigan U, 1987.

Brooks, Lisa. *Common Pot: The Recovery of Native Spaces in the Northeast.* Minneapolis: U of Minnesota P, 2008.

Brumble, H. David. *American Indian Autobiography*. Berkeley: U of California P, 1988.

Buell, Lawrence. *The Environmental Imagination: Thoreau, Nature Writing and the Formation of American Culture*. Cambridge: Belknap P, 1995.

Burnham, Michelle. *Captivity and Sentiment: Cultural Exchange in American Literature, 1682–1861*. Hanover, NH: UP of New England, 1997.

———. "'However Extravagant the Pretension': Bivocalism and US Nation-Building in *A Narrative of the Life of Mrs Mary Jemison*." *Nineteenth-Century Contexts* 23.3: 325–47.

Castiglia, Christopher. *Bound and Determined: Captivity, Culture-Crossing, and White Womanhood from Mary Rowlandson to Patty Hearst*. Chicago: U of Chicago P, 1996.

Clifton, James A. "Alternate Identities and Cultural Frontiers." *Being and Becoming Indian: Biographical Studies of North American Frontiers*. Ed. James A. Clifton. Chicago: The Dorsey P, 1989. 1–38.

Clements, William M. "Schoolcraft as Textmaker." *The Journal of American Folklore* 103.408 (April-June 1990): 177–92.

Deloria, Philip Joseph. *Playing Indian*. New Haven: Yale UP, 1998.

DeLucia, Christine. "Placing Joseph Bruchac: Native Literary Networks and Cultural Transmission in the Contemporary Northeast." *Studies in American Indian Literatures* 24.3 (Fall 2012): 71–96.

Derounian, Kathryn Zabelle. "Captivity Narrative in the Seventeenth Century." *Early American Literature* 23.3 (1988): 239–61.

Derounian-Stodola, Kathryn Zabelle. "The Captive as Celebrity." *Lives Out of Letters: Essays on American Literary Biography and Documentation, in Honor of Robert N. Hudspeth*. Ed. Robert D. Habich. Cranbury, NJ: Fairleigh Dickinson UP, 2004. 65–92.

———. *The War in Words: Reading the Dakota Conflict through the Captivity Literature*. Lincoln: U of Nebraska P, 2009.

Derounian-Stodola, Kathryn Zabelle and James Levernier. *The Indian Captivity Narrative, 1550–1900*. New York: Twayne, 1993.

Drake, Benjamin. *Life and Adventures of Black Hawk, with Sketches of Keokuk, the Sac and Fox Indians, and the Black Hawk War*. 1838. Cincinnati: George Conclin P, 1846.

Drake, Samuel Gardner. *The Book of the Indians of North America*. 2nd ed. Boston, Josiah Drake, 1835.

Ebersole, Gary L. *Captured by Texts: Puritan to Postmodern Images of Indian Captivity*. Charlottesville: UP of Virginia, 1995.

Erdrich, Louise. "Introduction." Tanner, John. *The Falcon: A Narrative of the Captivity and Adventures of John Tanner.* New York: Penguin Nature Library, 1994.

Ewan, Joseph. *Rocky Mountain Naturalists*. Denver: U of Denver P, 1950.

Faery, Rebecca Blevins. *Cartographies of Desire: Captivity, Race, and Sex in the Shaping of an American Nation*. Norman: U of Oklahoma P, 1999.

Faludi, Susan. *The Terror Dream: Fear and Fantasy in Post-9/11 America*. New York: Henry Holt, 2007.

Fierst, John T. "Return to 'Civilization'; John Tanner's Troubled Years at Sault Ste. Marie." *Minnesota History* 50.1 (1986): 23–36.

———. "A 'Succession of Little Occurrences': Scholarly Editing and the Organization of Time in John Tanner's *Narrative*." *Scholarly Editing: The Annual of the Association for Documentary Editing*. 33 (2012): 1–29.

———. "Strange Eloquence: Another Look at 'The Captivity and Adventures of John Tanner.'" *Reading beyond Words: Contexts for Native History*. Ed. Jennifer Brown and Elizabeth Vibert. Peterborough: Broadview P, 1996. 220–41.

Finseth, Ian Frederick. *Shades of Green: Visions of Nature in the Literature of American Slavery, 1770–1860*. Athens: U of Georgia P, 2009.

Franklin, Wayne. "Writing America from Abroad: Cooper's Recollected Sources in *The Wept of Wish-Ton-Wish*." *Literature in the Early American Republic*. 3 (2011): 1–39.

Frazee, George. "The Iowa Fugitive Slave Case." *Annals of Iowa* 4.2 (July 1899): 118–37.

Ganondagan. "Seneca Art and Culture Center." http://www.ganondagan.org/sacc.

Gianquitto, Tina. *"Good Observers of Nature": American Women and the Scientific Study of the Natural World, 1820–1885*. Athens: U of Georgia P, 2007.

Hausdoerffer, John. *Catlin's Lament: Indians, Manifest Destiny, and the Ethics of Nature*. Lawrence: U of Kansas P, 2009.

Helton, Tena L. "What the White 'Squaws' Want from Black Hawk: Gendering the Fan-Celebrity Relationship." *American Indian Quarterly* 34.4 (Fall 2010): 498–520.

"History of Watch Tower Park." Rock Island/Milan School District. Accessed 10 October 2009.

Howe, Susan. *The Birth-mark: Unsettling the Wilderness in American Literary History*. Hanover: Wesleyan UP, 1993.

Jackman, Sydney W. and John F. Freeman. "Editors' Introduction." *American Voyageur: The Journals of David Bates Douglass*. David Bates Douglass. Marquette: Northern Michigan UP, 1969. xiii-xxii.

Jackson, Donald. "Introduction." *Black Hawk, an Autobiography*. By Black Hawk. Urbana: U of Illinois P, 1955. 1–31.

Johnson, Kendall. "Peace, Friendship, and Financial Panic: Reading the Mark of Black Hawk in *Life of Ma-Ka-Tai-Me-She-Kia-Kiak*." *American Literary History*. Advance Access Publication (September 14, 2007): 771–99.

Johnson, Rochelle. *Passions for Nature: Nineteenth-Century America's Aesthetics of Alienation*. Athens: U of Georgia P, 2009.

Jordan III, William R. *The Sunflower Forest: Ecological Restoration and the New Communion with Nature*. Berkeley: U of California P, 2003.

Kane, Lucile M., June D. Holmquist and Carolyn Gilman. "Introduction." *The Northern Expeditions of Stephen H. Long: The Journals of 1817 and 1823 and Related Documents*. Stephen H. Long. St. Paul: Minnesota Historical Society, 1978.

Keller, Robert H. "Lac La Croix: Rumor, Rhetoric, and Reality in Indian Affairs." *The Canadian Journal of Native Studies* 8.1 (1988): 59–72.

Kilcup, Karen L. *Fallen Forests: Emotion, Embodiment, and Ethics in American Women's Environmental Writing, 1781–1924*. Athens: U of Georgia P, 2013.

Kolodny, Annette. *The Land before Her: Fantasy and Experience of the American Frontiers, 1630–1860*. Chapel Hill: U of North Carolina P, 1984.

Krupat, Arnold. *All That Remains: Varieties of Indigenous Expression*. Lincoln: U of Nebraska P, 2009.

———. *Ethnocriticism: Ethnography, History, Literature*. Berkeley: U of California P, 1992.

———. *For Those Who Come After: A Study of Native American Autobiography*. Berkeley: U of California P, 1985.

———. *"That the People Might Live": Loss and Renewal in Native American Elegy*. Ithaca, NY: Cornell UP, 2012.

Lepore, Jill. *The Name of War: King Philip's War and the Origins of American Identity*. New York: Knopf, 1998.

Lleweyllan, Carol White. "Portrait of G. Peter Jemison, the Artist." Ganondagan. http://www.ganondagan.org/Support/Friends-Of-Ganondagan/Peter-Jemison.

Logan, Lisa M. "'Cross-Cultural Conversations': The Captivity Narrative." *A Companion to Literatures of Colonial America*. Ed. Susan Castillo and Ivy Schweitzer. Malden: Blackwell Publishing, 2005. 464–79.

Lyndgaard, Kyhl. "Sarah Winnemucca Goes to Washington." *Nevada Historical Society Quarterly*. 58:1–4 (2015): 27–43.

Maddox, Lucy. *Removals: Nineteenth-Century American Literature and the Politics of Indian Affairs*. New York: Oxford UP, 1991.

Marsden, Michael T. "Henry Rowe Schoolcraft: A Reappraisal." *The Old Northwest* 2.2 (1976): 153–82.

Mason, Randall. "Historic Preservation, Public Memory, and the Making of Modern New York City." *Giving Preservation a History: Histories of Historic Preservation in the United States*. Ed. Max Page and Randall Mason. New York: Routledge, 2004. 131–62.

Mazel, David. *American Literary Environmentalism*. Athens: U of Georgia P, 2000.

Mielke, Laura L. *Moving Encounters: Sympathy and the Indian Question in Antebellum Literature*. Amherst: U of Massachusetts P, 2008.

Miles, George. "The Edwin James Letter Book at Yale." *Montana: The Magazine of Western History* 41.4 (Autumn 1991): 35.

Miranda, Deborah. *Indian Cartography*. Greenfield Center, NY: Greenfield Review P, 1999.

Mitchell, Lee Clark. *Witnesses to a Vanishing America: The Nineteenth-Century Response*. Princeton: Princeton UP, 1981.

Muller, Gilbert H. *William Cullen Bryant: Author of America*. Albany: SUNY P, 2008.

Myers, Jeffrey. *Converging Stories: Race, Ecology, and Environmental Justice in American Literature*. Athens: U of Georgia P, 2005.

Namias, June. "Editor's Introduction." *A Narrative of the Life of Mrs. Mary Jemison*. By James E. Seaver. Norman: U of Oklahoma P, 1992. 3–49.

———. *White Captives: Gender and Ethnicity on the American Frontier*. Chapel Hill: U of North Carolina P, 1993.

"Natural Features." Black Hawk State Historic Site. http://www.blackhawkpark.org/nf.htm. 19 March 2010.

Nichols, Roger L. "Introduction." *Black Hawk's Autobiography*. By Black Hawk. Ames: Iowa State UP, 1999. xi-xxi.

Pammel, L. H. "Dr. Edwin James." *Annals of Iowa* 8.3 (October 1907): 160–85; *Annals of Iowa* 8.4 (January 1908): 277–95.

Parker, Robert Dale. "Introduction: The World and Writing of Jane Johnston Schoolcraft." *The Sound the Stars Make Rushing Through the Sky: The Writings of Jane Johnston Schoolcraft*. By Jane Johnston Schoolcraft. Ed. Robert Dale Parker. Philadelphia: U of Pennsylvania P, 2006. 1–85.

Philippon, Daniel J. *Conserving Words: How American Nature Writers Shaped the Environmental Movement.* Athens: U of Georgia P, 2004.

———. "Introduction." Mabel Osgood Wright. *The Friendship of Nature: A New England Chronicle of Birds and Flowers.* Baltimore: The Johns Hopkins UP, 1999. 1–27.

Quaife, Milo Milton. "Introduction." *Life of Black Hawk, or Ma-Ka-Tai-Me-She-Kia-Kiak.* By Black Hawk. Chicago: R. R. Donnelley & Sons, 1916. 11–22.

Rifkin, Mark. "Documenting Tradition: Territoriality and Textuality in Black Hawk's Narrative." *American Literature* 80.4 (Dec. 2008): 677–705.

Rios, Theodore, and Kathleen Mullen Sands. *Telling a Good One: The Process of a Native American Collaborative Biography.* Lincoln: U of Nebraska P, 2000.

Rishell, James D. "Introduction." *Life of Black Hawk, or Ma-Ka-Tai-Me-She-Kia-Kiak.* By Black Hawk. Rock Island: American Publishing Co., 1912. viii–xi.

Rotella, Carlo. "Travels in a Subjective West: The Letters of Edwin James and Major Stephen Long's Scientific Expedition of 1819–1820." *Montana: The Magazine of Western History* 41.4 (Autumn 1991): 20–34.

Sac & Fox Nation. "Welcome to the Sac and Fox Nation." http://sacandfoxnation-nsn.gov/.

Salisbury, Neal. "Introduction: Mary Rowlandson and Her Removes." *The Sovereignty and Goodness of God, Together with the Faithfulness of His Promises Displayed, Being a Narrative of the Captivity and Restoration of Mrs. Mary Rowlandson and Related Documents.* By Mary Rowlandson. 1682. Boston: Bedford/St. Martin's, 1997. 1–60.

Sargent, Theodore D. *The Life of Elaine Goodale Eastman.* Lincoln: U of Nebraska P, 2005.

Sayre, Gordon M. "Abridging between Two Worlds: John Tanner as American Indian Autobiography." *American Literary History* 11.3 (Autumn 2001): 480–99.

———. *The Indian Chief as Tragic Hero: Native Resistance and the Literatures of America, from Moctezuma to Tecumseh.* Chapel Hill: U of North Carolina P, 2005.

———. "John Tanner, Métis: On the Impossibilities of Cultural Translation." *Native American Studies across Time and Space: Essays on the Indigenous Americas.* Ed. Oliver Scheiding. Heidelberg, Germany: Universitätsverlag Winter, 2010. 133–44.

Schmitz, Neil. *White Robe's Dilemma: Tribal History in American Literature.* Amherst: U of Massachusetts P, 2001.

Sellers, Stephanie A. *Native American Autobiography Redefined: A Handbook.* New York: Peter Lang, 2007.

Sewell, David R. "'So Unstable and Like Mad Men They Were': Language and Interpretation in Early American Captivity Narratives." *A Mixed Race: Ethnicity in Early America.* Ed. Frank Shuffelton. New York: Oxford UP, 1993. 39–55.

Shakopee Mdewakanton Sioux Community. "Environmental Restoration Efforts." Web.

"Shakopee Tribe in Minnesota Using Gambling Money to Restore Their Way of Life." Narr. Roger O'Neil. *NBC Nightly News.* 15 July 2008. Transcript.

Sharrock, Susan R. "Crees, Cree-Assiniboines, and Assiniboines: Interethnic Social Organization on the Far Northern Plains." *Ethnohistory* 21:2 (Spring 1974): 95–122.

Sherfy, Michael J. "Narrating Black Hawk: Indian Wars, Memory, and Midwestern Identity." Diss. U of Illinois at Urbana-Champaign, 2005.

Sivils, Matthew Wynn. *American Environmental Fiction, 1782–1847*. Surrey, UK: Ashgate, 2014.

Slotkin, Richard. *The Fatal Environment: The Myth of the Frontier in the Age of Industrialization, 1800–1890*. 1985. New York: Harper-Perennial, 1994.

Starre, Alexander. "(Forced) Walks on the Wild Side: Precarious Borders in American Captivity Narratives." *Ecozon@* 1.2 (2010): 22–37.

———. "Wilderness Woes: Negotiating Discourse and Environment in Early American Captivity Narratives." *Transcultural Spaces: Challenges of Urbanity, Ecology, and the Environment in the New Millenium*. Ed. Stefan Brandt and Frank Mehring. Tübingen: Narr, 2010. 273–93.

Steere, Judge Joseph H. "Sketch of John Tanner, Known as the 'White Indian.'" *Michigan Pioneer and Historical Collections* 22 (1894): 246–54.

Sweet, Timothy. "Masculinity and Self-Performance in the *Life of Black Hawk*." *American Literature* 65.3 (September 1993): 475–99.

Tawil, Ezra F. "Domestic Frontier Romance, or, How the Sentimental Heroine Became White." *NOVEL: A Forum on Fiction* 32.1 (Fall 1998): 99–124.

Tompkins, Jane. *Sensational Designs: The Cultural Work of American Fiction, 1790–1860*. New York: Oxford UP, 1985.

Trask, Kerry A. *Black Hawk: The Battle for the Heart of America*. New York: Henry Holt, 2006.

USDA, NRCS. 2009. The PLANTS Database profile for Cypripedium acaule (http:// plants.usda.gov /java/profile?symbol=CYAC3). 13 July 2009. National Plant Data Center, Baton Rouge, LA 70874–4490 USA.

Vizenor, Gerald. *Manifest Manners: Postindian Warriors of Survivance*. Hanover: Wesleyan UP, 1994.

Wallace, Mark. "Black Hawk's *An Autobiography*: the Production and Use of an 'Indian' Voice." *American Indian Quarterly* 18.4 (Fall 1994): 481–94.

Walsh, Susan. "'With Them Was My Home': Native American Autobiography and *A Narrative of the Life of Mrs. Mary Jemison*." *American Literature* 64.1 (March 1992): 49–70.

Weaver, Jace. *That the People Might Live: Native American Literatures and Native American Community*. New York: Oxford UP, 1997.

White, Daniel. "Antidote to Desecration: Leslie Marmon Silko's Nonfiction." *Leslie Marmon Silko: A Collection of Critical Essays*. Eds. Louise K. Barnett and James Thorson. Albuquerque: U of New Mexico P, 1999. 135–48.

Williams, Roger L. *'A Region of Astonishing Beauty': The Botanical Exploration of the Rocky Mountains*. Lanham: Roberts Rinehart, 2003.

Wyss, Hilary E. "Captivity and Conversion: William Apess, Mary Jemison, and Narratives of Racial Identity." *American Indian Quarterly* 23.3/4 (Summer/Fall 1999): 63–82.

Index